Water Security and the Food-Water-Energy Nexus: Drivers, Responses and Feedbacks at Local to Global Scales

ICCE Scientific Assembly 2017

Proceedings of the International Association of Hydrological Sciences (PIAHS) Volume 376, 2018

Port Elizabeth, South Africa
10-14 July 2017

Editors:

G. Jewitt
B. Croke

ISBN: 978-1-5108-5819-0

Printed from e-media with permission by:

Curran Associates, Inc.
57 Morehouse Lane
Red Hook, NY 12571

CURRAN ASSOCIATES INC. proceedings.com

Some format issues inherent in the e-media version may also appear in this print version.

Printed by Curran Associates, Inc. (2018)

International Association of Hydrological Sciences
Centre for Ecology and Hydrology
Wallingford, Oxfordshire OX10 8BB
United Kingdom

Phone: 441 491 692 442
Fax: 441 491 692 448

http://iahs.info/

Additional copies of this publication are available from:

Curran Associates, Inc.
57 Morehouse Lane
Red Hook, NY 12571 USA
Phone: 845-758-0400
Fax: 845-758-2633
Email: curran@proceedings.com
Web: www.proceedings.com

TABLE OF CONTENTS

Proc. IAHS, 376, 1–1, 2018
https://doi.org/10.5194/piahs-376-1-2018

Open Access PIAHS

Editorial: Special Issue on Water security and the food-water-energy nexus: drivers, responses and feedbacks at local to global scales

Barry Croke[1] **and Graham Jewitt**[2]

[1]Fenner School of Environment and Society, and Mathematical Sciences Institute, Australian National University, Australia
[2]Centre for Water Resources Research, University of KwaZulu-Natal, South Africa

Correspondence: Barry Croke (barry.croke@anu.edu.au)

Published: 1 February 2018

The papers presented in this special issue were part of a symposium held during the IAHS Scientific Assembly in Port Elizabeth, South Africa, in July 2017. Contributions were invited that investigated the issue of water security and more broadly the food-water-energy nexus; including investigation of water quality as well as quantity, hydro-economics, education, transboundary issues, the influence of processes operating at local to global scales, as well as those that compared conclusions drawn from local and global studies.

The papers presented in this special issue cover a wide range of topics, and have a broad geographical focus. Grigg et al. and Goomo et al. focus on the food-energy-water nexus, with Grigg et al. discussing the biophysical and socioeconomic challenges facing Pakistan, while Goomo et al. consider how to support better decisions, focusing on the Zambezi Basin. Some of the other papers focused on the water-energy part of the nexus (Cullis et al., Engström et al., and Adegun et al.). Cullis et al. present an approach to modelling the water-energy nexus to inform the planning process in the water and energy sectors. Engstrom et al. illustrate that local actions to mitigate climate change may significantly affect water resources elsewhere and Adegun et al., consider the interplay between water resources, hydropower generation and agricultural landuse.

The remaining papers have more of a focus on water security. Croke et al. focused on the impact of water and food security on livelihoods in rural India and Bangladesh; Xiaojun et al., Madrigal et al., Romashchenko et al., and Ghembreamlak et al. look at water demand and security in very different situations across the globe (China, Mexico, Ukraine and Sudan respectively); while Gusev et al., Kok et al., and Odunuga et al. focus more on agricultural systems. Gusev et al. report on modelling the effectiveness of mulch in improving soil moisture, crop yield and energy costs in Russia and Ukraine. Kok et al. discuss the potential of trade in phosphorus within Africa to address food security. Odunuga et al. discuss the potential of control measures in managing soil erosion in the Orashi River Basin in Nigeria. Finally, Yang et al. looked at the potential impact of future climate change on water resources.

Published by Copernicus Publications on behalf of the International Association of Hydrological Sciences.

Proc. IAHS, 376, 3–8, 2018
https://doi.org/10.5194/piahs-376-3-2018

Modelling the water energy nexus: should variability in water supply impact on decision making for future energy supply options?

James D. S. Cullis[1], Nicholas J. Walker[1], Fadiel Ahjum[2], and Diego Juan Rodriguez[3]

[1]Aurecon, Cape Town, 7441, South Africa
[2]Energy Research Centre, University of Cape Town, Cape Town, 7701, South Africa
[3]Water Global Practice, World Bank Group, Washington, DC 20433, USA

Correspondence: James D. S. Cullis (james.cullis@aurecongroup.com)

Received: 17 June 2017 – Accepted: 25 July 2017 – Published: 1 February 2018

Abstract. Many countries, like South Africa, Australia, India, China and the United States, are highly dependent on coal fired power stations for energy generation. These power stations require significant amounts of water, particularly when fitted with technology to reduce pollution and climate change impacts. As water resources come under stress it is important that spatial variability in water availability is taken into consideration for future energy planning particularly with regards to motivating for a switch from coal fired power stations to renewable technologies. This is particularly true in developing countries where there is a need for increased power production and associated increasing water demands for energy. Typically future energy supply options are modelled using a least cost optimization model such as TIMES that considers water supply as an input cost, but is generally constant for all technologies. Different energy technologies are located in different regions of the country with different levels of water availability and associated infrastructure development and supply costs. In this study we develop marginal cost curves for future water supply options in different regions of a country where different energy technologies are planned for development. These water supply cost curves are then used in an expanded version of the South Africa TIMES model called SATIM-W that explicitly models the water-energy nexus by taking into account the regional nature of water supply availability associated with different energy supply technologies. The results show a significant difference in the optimal future energy mix and in particular an increase in renewables and a demand for dry-cooling technologies that would not have been the case if the regional variability of water availability had not been taken into account. Choices in energy policy, such as the introduction of a carbon tax, will also significantly impact on future water resources, placing additional water demands in some regions and making water available for other users in other regions with a declining future energy demand. This study presents a methodology for modelling the water-energy nexus that could be used to inform the sustainable development planning process in the water and energy sectors for both developed and developing countries.

1 Introduction

International context and importance of the topic

Both energy and water are critical aspects of any economy, and yet despite their strong interdependence the two sectors are often managed independently (Hussey and Pittock, 2012). In resource use terms water is used to generate energy and energy is used to deliver, treat and supply water (Scott, 2011) (see Fig. 1) these links are referred as the water-energy nexus.

Developing an integrated approach to modelling the water-energy nexus is critical to supporting the development of effective national policies and regulations to ensure continued economic development and growth in a sustainable way (Bazillian et al., 2011).

Published by Copernicus Publications on behalf of the International Association of Hydrological Sciences.

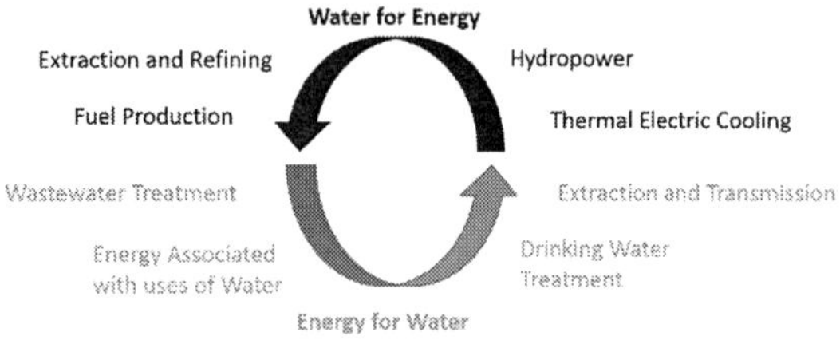

Figure 1. Defining the water energy nexus (after WEC, 2010).

Many countries, like South Africa, Australia, India, China and the United States, are highly dependent on coal fired power stations for energy generation. These power stations, however require significant amounts of water, particularly for cooling and when fitted with technologies to reduce pollution and climate change impacts.

Water availability is becoming more constrained in almost all countries through the combined effects of increasing demands, reducing water quality, and land use change. This presents a significant threat to future energy production (WEC, 2010). Similarly ever increasing water demands require consideration for more energy intensive technologies such as inter basin transfers (IBT), desalination and re-use of wastewater (Pittock, 2011).

As water resources come under ever increasing stress it is important that spatial variability in water availability is taken into consideration for future energy planning. Similarly, changes in energy policy are also likely to impact on water availability and also need to be taken into account for water resources planning and decision making.

This study seeks to make the link between energy and water resources policy and decision making through developing a linked water-energy nexus model. This study focuses on the water supply for energy generation part of the nexus. Specifically the objectives of this study are to incorporate regional variability in water availability and supply costs into an existing energy model so as to account for (1) a more representative cost of water supply to different technologies, (2) the spatial mismatch between water supply and the location of power plants, (3) the full cost of water supply to the energy sector including water supplied to mines, and (4) the opportunity costs of water use for energy production in a country with limited water resources and increasing demands.

South Africa was identified as a case study given its well documented water scarcity (DWAF, 2004), the importance of water for energy production, the extensive knowledge and strong analytical capacity for addressing the water-energy issue in the country, and the fact that the country is starting to plan water and energy in an integrated manner.

This South Africa case study is part of the World Bank's Thirsty Energy program and provides valuable knowledge and a framework for a water-energy nexus model that can be applied in other countries facing similar challenges relating to the water energy nexus (Rodríguez et al., 2013).

2 Water energy nexus in South Africa

2.1 South Africa's energy mix

Electricity supply in South Africa is dominated by a fleet of coal fired power stations operated by the state owned entity, Eskom. Eskom also functions as the system operator and owns and operates the transmission and distribution networks outside those owned and managed by the large metropoles. Eskom operates 22 power stations with a total nominal capacity of 41.9 GW, of which 85 % of the capacity is coal-fired. The balance of capacity is provided by nuclear, open-cycle gas turbine, hydro and pumped-storage power plants (ESKOM, 2013). In an attempt to address energy diversification, environmental concerns, and economic growth aspirations, energy sources such as nuclear, gas and renewables are being examined as alternatives by the Department of Energy (DoE). Of particular concern is the large volumes of water used for cooling of existing coal fired power stations and the benefits that renewable and alternative energy supply options might also provide in terms of improved water use efficiency in addition to reduced carbon impacts.

2.2 Water availability and supply

South Africa is a water-scarce country (annual freshwater availability is less than 1700 m^3 per capita), with limited average rainfall of about 450 mm yr^{-1} and unevenly distributed water resources (DWAF, 2004). South Africa has an annual mean-runoff value of only 40 mm per capita, one seventh of the global average of 260 mm, and rainfall and river flow are highly variable, erratic, and seasonal. Added to this is the fact that much of South Africa's key economic centres, including the urban and industrial centre of Gauteng, key mining areas and power stations, are located in areas of low water availability, far from major water sources where local demands exceed local supply. Out of necessity South Africa has had to develop a very proactive approach to water resources planning and supply which has resulted in a highly developed and integrated water supply system of large dams and many interbasin transfers to balance supply and demand. South Africa is the county with the 6th most number of large dams globally according to the International Commission of Large Dams (ICOLD), and with many kilometres of pipelines including large inter-basin transfer schemes (DWAF, 2004). Many of these have been developed primarily to provide water to coal fired power stations located in the Vaal, Mokolo, Olifants and Inkomati catchments. Nationally only about 2 % of water is used for power production (DWA, 2013a), but in some catchments this can be as high as 40 % and at a very high level of assurance.

Table 1. Regional variability in potential power generation activities in each water supply region and associated WMAs.

WSR	WMA	Region	Activity
A	Limpopo	Waterberg (Lephalale)	Open-cast coal mining Coal thermal power plants with FGD option Coal-to-Liquids refineries
B	Olifants	Mpumalanga, Witbank	Open-cast and underground coal mining Coal thermal power plants with FGD option Coal-to-Liquids refineries
C	Upper Vaal	Mpumalanga, Secunda	Open-cast and underground coal mining Coal thermal power plants with FGD option Inland gas thermal power plants Inland Gas-to-Liquids refineries
D1	Lower Orange	Northern Cape, Upington	Concentrated Solar Thermal Power Plants (CSP)
D2	Lower and Upper Orange	Northern Cape, Karoo	Shale gas mining Gas thermal power plants Inland gas-to-liquids refineries

2.3 Water supply and energy planning

The long-term infrastructure planning process for the supply of both energy and water is well established through government departments. The planning of both resources has taken into account cost and scarcity of the other to various degrees, but to date integrated modelling of the bulk supply infrastructure of both systems has not been undertaken. Investigating the significance of these linkages and how they affect future water and energy planning requires the integration of water constraints into energy models and energy constraints into water supply models.

3 The water energy nexus model

3.1 SATIM-W: A Water Energy Nexus Model

The South African TIMES model (SATIM), a public domain energy systems model developed by the University of Cape Town's Energy Research Centre (ERC), is a suitable base model for integrated water-energy planning[1]. SATIM is a national energy system model built using the TIMES model generator, which was developed under the auspices of the International Energy Agency's Energy Technology Systems Analysis Program (IEA-ETSAP).

Typically future energy supply options are modelled using an optimization model such as SATIM that considers water supply as an input costs, but is generally constant for all technologies. It is however important to note that different technologies are located in different regions of the country with varying levels of water availability and associated infrastructure development and supply costs. In order to address this limitation regional variability was introduced in to SATIM-W and individual water supply options, include major investments in dams and transfer projects and water supply energy needs, were incorporated into the SATIM-W model so as to capture the water-energy interplay. Incorporating a regional cost and quality for water allows the model to examine potential trade-offs within the supply sector e.g. fuel extraction and processing, treatment of water, cleaning and flue gas desulphurization. Regional variability in the non-energy water demands were also included in the model as this affects the relative opportunity cost for water supply to energy in each area. The updated model, SATIM-W, allows these activities to be represented so that the model is responsive to the regional cost and availability of water and energy supply, connected to a single national demand-side representation.

3.2 Water and energy production regions

South Africa is divided into 19 Water Management Areas (WMAs). The location of these WMAs, relative to the different energy producing regions of South Africa are shown in Fig. 2. These regions include: A: Waterberg (Lephalale); B: Mpumalanga, Witbank; C: Mpumalanga, Secunda; D1: Northern Cape, Upington; D2: Northern Cape, Karoo (Fig. 2). The nature of the specific energy activity in each of these regions is given in Table 1. In each region the Department of Water and Sanitation (DWS), formally known as the Department of Water Affairs (DWA) has assessed all future water supply options and determined the available yield, estimated capital and operational costs, and average energy requirements (e.g. for the pumping of water or desalination or water treatment) (DWA, 2010a).

[1]More detailed documentation of SATIM can be found at: http://www.erc.uct.ac.za/Research/esystems-group-satim.htm

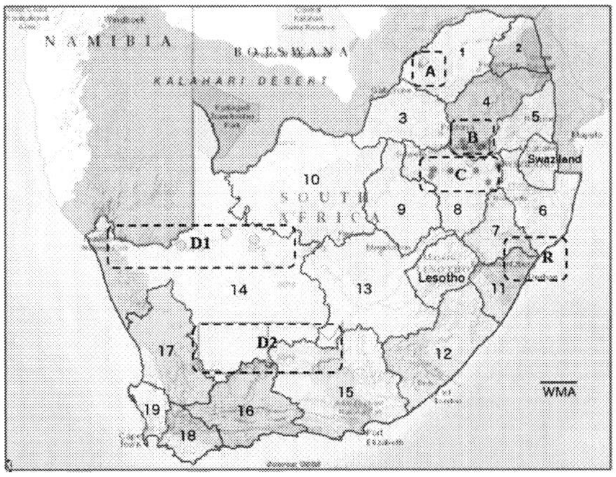

Figure 2. South African Water Management Areas and Energy Producing Regions (A: Waterberg (Lephalale); B: Mpumalanga, Witbank; C: Mpumalanga, Secunda; D1: Northern Cape, Upington; D2: Northern Cape, Karoo. The WMAs are numbers as per the original definition given in the National Water Resources Strategy (DWAF, 2004).

The regional distribution of water sources and consumers varies greatly in South Africa and as a result, the demand for water and supply is highly regionalized. For example, in the Waterberg district municipality (Lephalale) in Limpopo province, where the Waterberg coal deposits occur, the demand for water is dominated by water needs of the dry-cooled Matimba coal-fired power station for make-up water and other plant needs (4.3 million m^3 p.a.), and the water requirements of its supplier, the Grootgeluk coal mine, which use water for coal washing (9.9 million m^3 p.a.). Together these two demands account for approximately 40 % of the existing water withdrawals in the district. Energy sector withdrawals may grow to 75 % by the year 2030 if further developments in coal-based energy supply are pursued (Aurecon, 2014; van Vuuren, 2006). Approximately 20 % of current water withdrawals in the Waterberg region are directly attributed to electricity generation. This exceeds the availability of water in the local catchment area and has resulted in the construction of a major pipeline to transfer water in to the region from the neighbouring Vaal river system (Aurecon, 2011).

3.3 Regional water supply cost curves

A representative cost for water supply to the different energy producing regions was determined according to the *Revised Water Pricing Strategy for Raw Water* (DWA, 2012) and by using data provided in the analysis of the ultimate marginal cost of water supply to different regions in South Africa (DWA, 2010a). The unit cost of water supply for energy consists of the cost of the bulk infrastructure, delivery (transmission and distribution) and treatment requirements. The capital, fixed and variable operating and maintenance components are calculated separately in each water supply region for each water supply scheme (e.g., dam, inter-basin transfer) as part of determining the potential regional water supply cost. Where possible these costs have been updated with more recent cost estimates for specific schemes and regions (Aurecon, 2011; DWA, 2009, 2010b, 2013).

The marginal cost for water supply also increases with increasing demands and hence it was important to consider the impact of other non-energy water demands in each region and to develop a regional marginal cost curve for water supply as a function of the total demand in the region. This was also important in that the cost of water could also vary depending on when the proposed energy developments take place as future water supply options are generally more expensive than the current supply costs.

The resulting regional marginal cost curves for water supply are then included in the model associated with each region and SATIM-W then weights each water supply and delivery option (or scheme) and chooses the combination that delivers the needed water at least-cost, resulting in the determination of the total marginal water supply cost.

The costs for delivery of water to power plants is based upon estimates for deploying and managing major water supply and transfer schemes, but does not capture final details (and associated costs) that can only be determined when a specific site is identified. This is also true for hydraulic fracturing and Concentrated Solar Power where the exact locations and method of water delivery have not been determined. But in both cases these are rather small compared to the other costs characterizing each scheme.

4 Model scenarios and results

In order to assess the benefits of integrating the regional variability of water supply costs into future energy supply planning, two model scenarios were considered. The first scenario considered water supply as a uniform cost applied to all future energy supply options. The second scenario included the regional marginal costs of water supply as well as the other non-energy water demands relative to the locations in which the different future energy supply options where located. The results of these scenarios are given in Figs. 3 and 4 respectively. In order to model the potential impacts of climate change policy on future energy and water supply, a number of other scenarios were considered, including a limit on the total production of CO_2, as well as a dryer future scenario due to climate change. The results are shown in Fig. 5.

5 Discussion

In the absence of taking into account the regional variability of water supply and associated infrastructure costs (the No

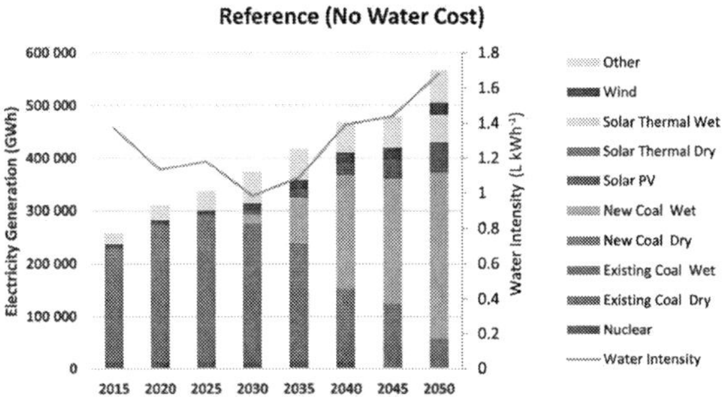

Figure 3. Optimal future energy mix under the reference scenario with no regional variability in water supply costs.

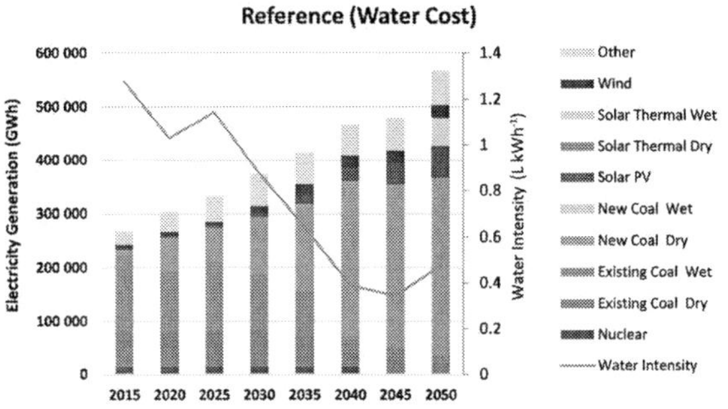

Figure 4. Optimal future energy mix under the reference scenario but with regional variability in water supply costs.

Water Cost scenario – Fig. 3), wet-cooled coal power plants are the preferred choice due to their lower investment costs and higher net generation efficiencies. However, when consideration is given to the regional variability in water supply (the Water Cost scenario in Fig. 4), dry-cooling is the preferred option for new coal power plants, particularly in the Waterberg region where the remaining economically viable coal reserves are located. New dry-cooled capacity of approximately 40 GW is commissioned by 2050 and includes the replacement of the existing stock of 37 GW which will mostly be retired by then. This SATIM-W result indicates that Eskom's dry cooling policy is really in the economic interests of the country, even though it increases the cost of electricity from coal power plants. This has a significant impact in terms of future water use efficiency in the power sector, which could either reach a peak of $1.65\,\mathrm{L\,kWh^{-1}}$ by 2050 based on the No Water Cost scenario, or $0.5\,\mathrm{L\,kWh^{-1}}$ based on the Water Cost scenario. In absolute terms, the inclusion of regional water supply cost cuts the cumulative (2010–2050) water consumption for the power sector by 9338 million m^3 (77 %) with just a modest increase (0.84 %) in the system cost.

Other than water consumed by power plants, the two scenarios have similar total system cost, energy supply expenditures, and primary and final energy consumption. Interestingly, the Water Cost scenario produces slightly more CO_2 emissions in spite of generating 1.3 % less electricity with coal and 2 % more with RE technologies (chiefly wind and solar PV, which require no water to generate electricity). This results from the reduced efficiency and higher unit emissions that are associated with the dry-cooled coal plants that are adopted when water costs are taken into account.

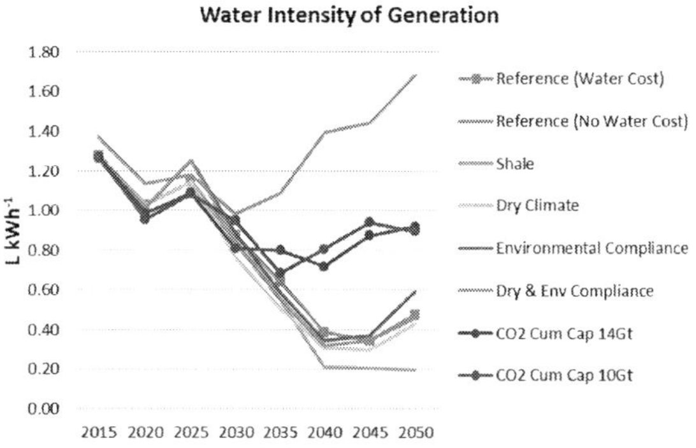

Figure 5. Resulting power sector water consumption and efficiency ($\mathrm{l\,kWh^{-1}}$) under different model scenarios.

The impact of climate change policy, such as imposing a carbon limit on the energy sector, results in a significant shift in the optimal future energy mix. In particular this results in increase demand for renewables, but particularly concentrated solar power (CSP). The result of this is surprisingly an increase in the overall water use efficiency for energy production as shown in Fig. 5 as compared to the reference case Water Cost scenario which is dominated by dry-cooled coal fired power stations. The imposing of a carbon cap also has a significant impact on future water resources planning as it looks to develop more CSP plants, which are located in along the Orange River at the expensive of existing coal fired power stations in the Olifants and Upper Vaal catchments or new coal in the Lephalele area. This could result in a shift in water demand leading not only to stranded assets in terms of decommission power stations, but also the existing bulk water distribution systems that have been developed over the year to supply these power stations with water.

6 Conclusions and recommendations

These findings illustrate the insights gained from integrated and regionally disaggregated water-energy modelling and in particular the importance of taking in to consideration the regional variability in water supply costs and associated locations for different energy technologies. The main message from this study is that water and energy planning should be integrated. The analysis also shows the ability to identify major water infrastructure investments that could become stranded (or sub-optimal from an economic perspective) in light of possible future energy policy changes and can thereby help to formulate hedging strategies aimed at minimizing the likelihood of these stranded assets. Alternatively however, the opportunity to utilise these "stranded" water infrastructure investments could represent a significant economic opportunity as these could be repurposed to provide water needed to support other activities in the region at a relatively low costs. Hence it is important not only that water and energy policy be considered together in the context of the water-energy nexus, but that this should also be integrated with regional economic development and planning particularly in the context of future climate change and other uncertainties. This study also does not include additional externalities such as impacts on water quality or the total impact on CO_2 emissions. These can however be incorporated into the further development of the model and future research.

Data availability. UCT, the custodians of the model and model results can be accessed: http://energydata.uct.ac.za/.

Competing interests. The authors declare that they have no conflict of interest.

Special issue statement. This article is part of the special issue "Water security and the food–water–energy nexus: drivers, responses and feedbacks at local to global scales". It is a result of the IAHS Scientific Assembly 2017, Port Elizabeth, South Africa, 10–14 July 2017.

Acknowledgements. This work was made possible by the financial contribution of the Water Partnership Program of the Water Global Practice, World Bank Group which supports the Thirsty Energy Initiative The authors would also like acknowledge the support of the members of the Thirsty Energy program steering committee as well as the anonymous reviewers for providing additional guidance.

Edited by: Barry Croke
Reviewed by: Fortune Faith Gomo and one anonymous referee

References

Aurecon: Development of a reconciliation strategy for the Olifants River Water Supply System: water requirements and water resources report, Report to Department of Water Affairs, Pretoria, RSA, report no.: P WMA 04/B50/00/8310/6, 2011.

Aurecon: Modelling the Water Energy Nexus in South Africa Task 1 Report: Development of Regional Marginal Water Supply Cost Curves, Document prepared by: James Cullis and Nicholas Walker, Aurecon South Africa (Pty) LTD, for the World Bank Thirsty Energy Program, 22 September 2014, 2014.

Bazilian, M., Rogner, H., Howells, M., Hermann, S., Arent, D., Gielen, D., Steduto, P., Mueller, A., Komor, P., Tol, R. S. J., and Yumkella, K. K.: Considering the energy, water and food nexus: Towards an integrated modelling approach, Energ. Policy, 39, 7896–7906, doi:10.1016/j.enpol.2011.09.039, 2011.

Department of Water Affairs (DWA): Development of an Integrated Water Quality Management Plan for the Vaal River System Task 5: Water Quality Economic Impact Modelling Final Report, Department of Water Affairs & Forestry, Pretoria, South Africa, Report No. PRSA C000/00/2305/4, September 2009.

Department of Water Affairs (DWA): Assessment of the ultimate potential and future marginal cost of water resources in South Africa DWA Report No. P RSA 000/00/125610, Department of Water Affairs, Pretoria, South Africa, September 2010a.

Department of Water Affairs (DWA), Integrated Water Resource Planning for South Africa: A Situation Analysis, Report Number: P RSA 000/00/12910, Department of Water Affairs, Pretoria, South Africa, 2010b.

Department of Water Affairs (DWA), Revised Water Pricing Strategy for Raw Water III, Draft for comment, Department of Water Affairs, Pretoria, South Africa, available at: www.dwa.org.za, 2012.

Hussey, K. and Pittock, J.: The energy-water nexus: managing the links between energy and water for a sustainable future, Ecol. Soc., 17, 31, doi:10.5751/ES-04641-170131, 2012.

Pittock, J.: National climate change policies and sustainable water management: conflicts and synergies, Ecol. Soc., 16, 25, 2011.

Scott, C. A., Pierce, S. A., Pasqualetti, M. J., Jones, A. L. Montz, B. E., and Hoover, J. H.: Policy and institutional dimensions of the water-energy nexus, Energ, Policy, 39, 6622–6630, doi:10.1016/j.enpol.2011.08.013, 2011.

Proc. IAHS, 376, 9–13, 2018
https://doi.org/10.5194/piahs-376-9-2018

The water–food–energy nexus in Pakistan: a biophysical and socio-economic challenge

Nicola Grigg[1], Tira Foran[1], Toni Darbas[2], Mac Kirby[1], Matthew J. Colloff[3], Mobin-ud-Din Ahmad[1], and Geoff Podger[1]

[1]CSIRO Land and Water, ACT 2600 Canberra, Australia
[2]CSIRO Land and Water, Brisbane, 4001 Queensland, Australia
[3]Fenner School of Environment & Society, Australian National University, ACT 2601 Canberra, Australia

Correspondence: Nicola Grigg (nicky.grigg@csiro.au)

Received: 16 June 2017 – Accepted: 25 July 2017 – Published: 1 February 2018

Abstract. We draw on previous work examining historical trends, likely future water use and food availability in Pakistan and extend the analysis to consider interactions with hydropower generation and the energy demand in food production due to pumping of groundwater for irrigation. Business-as-usual scenarios suggest growing demands for groundwater and energy use for food production as population grows rapidly. However, groundwater use is already unsustainable in many areas, and energy supply is failing to keep up with demand. Quantifying material linkages between water, food and energy provides a means to explore biophysical constraints. Characterising institutional constraints is equally important, as they can be significant barriers to effective stewardship of water, energy and food resources. The experience in Pakistan reinforces this finding, and we discuss the implications for hydrologists.

1 Background

The water–energy–food nexus is conceptualised and analysed in many ways, with various authors offering generalizable insights. We consider some of these insights in relation to the Indus Basin in Pakistan.

Scott et al. (2015) considered that "resource recovery is the basic biophysical expression of the nexus approach", and that biophysical limits ultimately require resource uses that do not externalise unwanted impacts. The authors emphasised resource linkages beyond a biophysical analysis to include materials, institutions and security.

Kirby et al. (2017) characterised the material linkages between water and food for three population growth scenarios in Pakistan, with a focus on one aspect of food security, food availability, and provided biophysical bounds and possibilities for the water-food system by exploring water use and crop yield assumptions. In this paper we extend the analysis to include material interactions with energy and, following Scott et al. (2015), anticipate beyond material linkages to consider institutional and security dimensions. "Security" here refers to "human and ecosystem dependence dimensions of resource security". We include considerations of resilience and adaptive capacity.

2 The water–energy–food nexus in Pakistan

The population estimate for Pakistan is almost 190 million. From 1961 to 2013, population grew 3.8 times (UN, 2015), whereas grain production grew 5.2 times. Food availability has improved, but many people remain malnourished (Kirby et al., 2017). Increased food production was achieved by increased use of groundwater for irrigation, while diversions of surface water have remained unchanged. In 2007, about 77 billion cubic metres (bcm) of water was supplied from rivers, and 40 bcm from groundwater (Kirby et al., 2017). It is generally agreed that groundwater is overused in many areas, particularly in the Punjab, and continued use at this rate is unsustainable (Kirby et al., 2017) and risks non-recoverable expansion of saline areas.

Farm energy use is increasing faster than crop production. Groundwater pumping (mostly diesel) represents 60 % of agricultural energy demand (Siddiqi and Wescoat Jr., 2013).

Table 1. Material consequences of five categories of food-water activities and two energy options. Consequences are relative to Kirby et al. (2017) "yield only" scenario with median population projection.

Approach	Water perspective	Energy perspective	Food perspective
(1) Increase crop area and water use	Risk of higher water demand and of demand exceeding supply. Potential for irrigation demands to be affected by hydropower requirements.	Agricultural demand for energy is low relative to other sectors, but hydropower vs. irrigation tensions could be heightened under this option.	Higher crop production, with higher dependence on affordable and reliable water and energy supply.
(2) Increase yield per unit water; or (3) swap to lower water use crops	Potential to lower agricultural water demand, but only if system rebounds are managed.	Potential to lower agricultural dependence on energy supply, but only if system rebounds are managed.	Potential for same production with lower dependence on energy and water.
(4) Import more food, export less; or (5) slower population growth	Lower demand for irrigation water.	Lower agricultural demand for energy, and potential to ease tensions between hydropower and irrigation requirements.	Potential to ease demand for food from agricultural sector.
(6) More hydropower from (a) large dams and/or (b) local barrages	May help or hinder provision of irrigation water depending on dam locations and operations.	Improves energy supply. Option (b) may help build local supplies without large investment in new grid infrastructure.	Linked to water. May ease other barriers to food access (e.g. opening new livelihood options, more reliable cold storage).
(7) Import more energy	Relieves constraints that hydro places on irrigation water.	Increases dependence on affordable and reliable energy markets.	Linked to water. May ease other barriers to food access (e.g. opening new livelihood options, more reliable cold storage).

Although agricultural use of energy is only 2 % of total energy use (based on 2012–2013 figures from Ahmed et al., 2015), it contributes to demand for imported oil, exacerbating Pakistan's trade imbalance. The building of dams for irrigation increased availability of hydropower, but capacity and output have barely increased this century and electricity supply is regarded as in crisis (Energy Sector Task Force, 2010; Kugelman, 2015).

Population growth may reach 310 million by 2050 (UN, 2015), with associated demand for more food, water and energy. Kirby et al. (2017) suggested maintenance of adequate food availability will be via five approaches: (1) increased area and water use of food crops, requiring more water from over-used aquifers or more surface water storage; (2) increase in crop yields per unit of water (at greater rates than in the past); (3) swapping to crops with lower water demand; (4) exporting less food and importing more; and (5) slowing population growth, though this is likely to have only long-term impact.

These approaches represent statements of material assumptions rather than policy options: if by 2050, current food availability has been maintained, it will be through a combination of those five approaches, whether or not through planning and deliberation.

Decisions to increase energy supply can be made independently of water-food options, though there are clear interactions. Two energy options are: (6) build dams and hydroplants to supply more power (noting potential hydropower capacity is > 54 000 MW, compared to the current 6500 MW; Energy Sector Task Force, 2010), and (7) importing energy (Sharma et al., 2004). Option (6) could come from a few large dams, requiring major transmission infrastructure, or from local distributed sources such as small scale hydropower on irrigation canals and at other barrages which are close to demand centres (Energy Sector Task Force, 2010). The electricity grid needs significant upgrading to meet existing demand, let alone to accommodate new hydro or transboundary supplies which bring extra challenges in financing, locating, building and maintaining very long transmission lines in difficult terrain.

Material nexus perspectives on water-food options

At 2 % of total demand, there is little risk of agricultural energy demand impacting on national energy availability or prices. But high energy dependence means the agricultural sector is heavily impacted by energy prices, availability and reliability. In relation to hydropower and groundwater pumping, the five approaches of Kirby et al. (2017) have clear material implications for water, energy and food. Table 1 shows these impacts relative to the "yield only" scenario, where crop yields increase according to historical trends, but areas do not change. The first approach tightens coupling in the water–energy–food nexus by placing more demands on limited water resources. Crop production will have higher energy demand for groundwater pumping, and river operations will need to balance hydropower and irrigation requirements.

Approaches 2 and 3 can save water by lowering demand by efficient use (via irrigation practices and technology, improved yields and growing crops with low water-demand). Ahmad et al. (2014) found savings from efficiency measures at the farm-scale did not translate to savings at the basin-scale. One reason for this is because of system "rebound": local efficiency gains are consumed in other ways, leading to no system-scale saving (Scott et al., 2014). For example, increased efficiency enables farmers to increase land area under production for the same water use. In the absence of a basin-level cap on diversions, basin-level water use would increase until biophysical limits are reached. Thus approaches 2 and 3 would require political agreement, regulation and compliance for basin-scale diversion limits.

Approaches 4 and 5 could relieve tensions by reducing demand for food, energy and water, but there are limitations. For example, approach 4 implies structural transformation of the economy, conflicts with political-cultural preferences, and financing options are unclear. For approach 5, the population would continue to grow even if there was a large decrease in fertility now. These two approaches are not in the mandate of authorities responsible for water, energy and agriculture.

3 Potential nexus solutions

Nexus solutions have focused on large infrastructure such as hydropower dams and new irrigation schemes. Three large hydropower dams each with over 1000 MW installed capacity currently supply 80 % or more of Pakistan's hydropower of about 7000 MW (Private Power and Infrastructure Board, 2011). There are approximately 60 irrigation schemes on the Indus and its tributaries, with a total command area of about 18 million ha, making it the world's largest contiguous irrigation area (Archer et al., 2010). However, such projects face major challenges. International donors are reluctant to finance dams and storages designed only for irrigation. Dam projects are proposed in provinces of Gilgit-Baltistan and Azad Jammu Kashmir, in territory disputed with India, and require agreement from India to arrange international financing. The Kalabagh dam, not in disputed territory and which banks are willing to finance, is not supported by the government of Sindh province.

Large-scale irrigation efficiency is called for as a solution via delivery of easily-realised, readily acceptable gains. But, as discussed above, net efficiencies may be much lower than anticipated (Ahmad et al., 2014; Scott et al., 2014). Other, local, solutions may be politically more achievable: small-scale hydropower at head works; improved irrigation efficiency in well-identified places; improved water delivery; and managed aquifer storage to keep groundwater tables higher and reduce energy use in pumping. The private sector has an important role in small hydropower, as evidenced by projects in Khyber Pakhtunkhwa and Gilgit-Baltistan. However, projects on headwork/canal networks require endorsement by provincial government irrigation departments.

Our overview has been framed by the biophysical limits assessed by Kirby et al. (2017). However, it is clear the water–energy–food nexus is also characterised by institutional limits, which manifest differently according to the five approaches (Table 1). These include operational limits to manage a tightly-coupled nexus that maximises water, energy and food outputs, and limited socio-political capacity to create and implement basin-level solutions beyond the mandate of individual energy, water and agricultural authorities.

Such socio-political factors tend to be considered out of scope for biophysical scientists, other than in calling for "good resource governance". By so doing, biophysical scientists risk distancing themselves from implementable solutions that take them beyond specific domain knowledge and expertise, or the mandate of their organisation.

4 Alternative nexus perspectives

Mindful of this risk, we review nexus perspectives that include institutional and security considerations as a step towards addressing the need "to acknowledge the existence and legitimacy of a range of narratives and frames in pursuing a nexus approach" (Middleton et al., 2015).

One difference between nexus approaches is in how they treat agency. A resource efficiency approach, as applied by Kirby et al. (2017) and Laghari et al. (2012), presents technical means to maximise productivity within biophysical limits, but is silent on issues of actors and their agency (other than perhaps an unspoken assumption that compelling resource analyses will motivate government policy for change). Resource efficiency approaches acknowledge the need for policy and practice reform but focus on technical solutions to increase synergies and reduce trade-offs between supply and demand relating to water, food, energy and environment.

A critical social science perspective focuses on the power relations and societal structures that shape and reinforce resource-access tensions. Akhter (2017) considered "access to land … is arguably a much more relevant factor for evaluating water and food security than the total amount of water available", referencing estimates that 2 % of householders in the Indus Basin own almost half the land area. Social relations are key in shaping resource access options for different social groups, as evidenced by steep increase in rural landlessness associated with unequal distribution of the benefits of increased agricultural production: an outcome of the Green Revolution was concentrated land ownership for the few and reduced incomes and livelihood security for a significant proportion of the population. A critical social science perspective emphasises the need to reform deep inequities that represent limits as least as powerful as biophysical constraints.

Another way nexus approaches differ is in stakeholder engagement. A spectrum exists from participatory, multi-scale, multi-stakeholder resource management (commonly recommended, but rarely realised) to restricted participation by actors from state agencies with formal responsibilities for water, energy and agricultural sector planning and service provision. A third way nexus approaches differ is how questions of resilience, adaptation and transformation are addressed. At one end of the spectrum is maximising resource efficiencies and production; at the other efficiency and production are considered in light of natural disasters, system tipping points, rapid global change and shocks. Allouche et al. (2014) found current dominant nexus solutions were framed by "stability" and "durability", even though "clumsy solutions" involving exchange of differing world-views and problem perceptions were more appropriate for water–energy–food nexus dynamics.

For example, small-scale hydropower will not "solve" water–energy–food challenges in Pakistan, but have desirable attributes: they use existing barrages; do not require large investments in long-distance transmission lines; are well suited to limited electricity grids and raise fewer social-political concerns than large hydro projects. Two of the authors (GP and MA) developed a river system model for the Indus with multiple stakeholders familiar with complex river operations and the social-political decision context. By learning about the potential for small hydropower developments, new questions were addressed with the river system model, and new criteria were developed against which model scenarios could be assessed. Early indications are that including energy considerations brings potential benefits to water users. Barrages need maintenance that would be hard to fund if benefits were only to irrigation. A stronger case can be made if energy supply options are included, as well as safety and livelihood impacts of infrastructure failure.

This example is consistent with key messages of Allouche et al. (2014, 2015) that emphasise recognition of institutional limits, working with diverse kinds of storage and using pluralism to achieve "many ten percent" solutions. These may not be optimal from any single perspective, but are judged as acceptable according to a diverse range of perspectives and are thus more practical and effective than other solutions.

5 Conclusion

In the Indus basin, the biophysical limits of the water–energy–food nexus are just one piece of the picture. Characterising the material linkages is helpful for diagnosis, and for developing a testable evidence base for assumptions and hypotheses about the material consequences of different interventions. These approaches are useful because they provide insights independent of social and historical context to reveal commonalities that stakeholders can agree on, despite other differences. This approach was vital to the successful negotiation of the Indus Waters Treaty between Pakistan and India (Gulhati, 1973).

It is valuable to recognise, however, that in the Indus basin and elsewhere the water–energy–food nexus can be understood through critical social science and resilience or adaptation-oriented perspectives. Rather than alternative perspectives being at odds with biophysical analyses, or imposing constraints on solutions, these approaches can expand options usually considered out of scope for biophysical scientists or water, energy and agricultural authorities. For biophysical scientists engaging with the nexus, a willingness to include diverse perspectives presents new opportunities. On a technical level, it can reveal new questions to ask of biophysical models, and inform exploration of model scenarios to make analyses relevant to a wider social-political context.

Respectful humility and awareness of the limits of our knowledge helps bound us to our disciplinary training and experience. Being open to alternative perspectives requires suspension of judgment and time devoted to seeing a familiar situation with genuinely new eyes. Such re-framing is challenging because it represents a different way of engaging with the world and embracing uncertainty, ambiguity and scope-creep of dynamic, distributed "clumsy solutions" rather than pursuit of certainty and control. The capacity to re-frame also requires alternative perspectives to be held in juxtaposition to uncover insightful intersections or conflicts between them (Roe, 2012; Foran et al., 2014). In Pakistan the Vision 2025 Agenda and the Government of Pakistan's commitment to the Sustainable Development Goals could provide potential platforms that would support a broader range of perspectives than usually possible within sector-bound agencies.

Data availability. No data sets were used in this article.

Competing interests. The authors declare that they have no conflict of interest.

Special issue statement. This article is part of the special issue "Water security and the food–water–energy nexus: drivers, responses and feedbacks at local to global scales". It is a result of the IAHS Scientific Assembly 2017, Port Elizabeth, South Africa, 10–14 July 2017.

Acknowledgements. We thank Glen Currie (University of Melbourne) for discussions on the potential for small hydropower developments in Pakistan, and Deborah O'Connell and Peter Wallbrink (CSIRO) for providing helpful reviews.

Edited by: Barry Croke
Reviewed by: Fortune Faith Gomo and one anonymous referee

References

Ahmad, M.-D., Masih, I., and Giordano, M.: Constraints and opportunities for water savings and increasing productivity through Resource Conservation Technologies in Pakistan, Agr. Ecosyst. Environ., 187, 106–115, https://doi.org/10.1016/j.agee.2013.07.003, 2014.

Ahmed, M., Riaz, K., Maqbool Khan, A., and Bibi, S.: Energy consumption–economic growth nexus for Pakistan: Taming the untamed, Renew. Sustain. Energ. Rev., 52, 890–896, https://doi.org/10.1016/j.rser.2015.07.063, 2015.

Akhter, M.: The political ecology of the water scarcity/security nexus in the Indus Basin: decentering per capita water supply, in: Imagining Industan: Overcoming Water Insecurity in the Indus Basin, edited by: Adeel, Z. and Wirsing, R. G., Springer, Cham, 2–33, https://doi.org/10.1007/978-3-319-32845-4_2, 2017.

Allouche, J., Middleton, C., and Gyawali, D.: Nexus Nirvana or Nexus Nullity? A dynamic approach to security and sustainability in the water–energy–food nexus, STEPS Working Paper 63, STEPS Centre, Brighton, 2014.

Allouche, J., Middleton, C., and Gyawali, D.: Technical Veil, Hidden Politics: Interrogating the Power Linkages behind the Nexus, Water Altern., 8, 610–626, 2015.

Archer, D. R., Forsythe, N., Fowler, H. J., and Shah, S. M.: Sustainability of water resources management in the Indus Basin under changing climatic and socio economic conditions. Hydrol. Earth Syst. Sci., 14, 1669–1680, https://doi.org/10.5194/hess-14-1669-2010, 2010.

Energy Sector Task Force: Integrated energy sector recovery report and plan, Asian Development Bank, Manila, https://www.adb.org/publications/integrated-energy-sector-recovery-report-and-plan (last access: 1 September 2017), 2010.

Foran, T., Butler, J. R. A., Williams, L. J., Wanjura, W. J., Hall, A., Carter, L., and Carberry, P. S.: Taking complexity in food systems seriously: an interdisciplinary analysis, World Dev., 61, 85–101, https://doi.org/10.1016/j.worlddev.2014.03.023, 2014.

Gulhati, N. D.: Indus Waters Treaty: an exercise in international mediation, Allied Publishers, Bombay, 1973.

Kirby, M., Ahmad, M.-D., Mainuddin, M., Khaliq, T., and Cheema, M. J. M.: Agricultural production, water use and food availability in Pakistan: Historical trends, and projections to 2050, Agr. Water. Manage., 179, 34–46, https://doi.org/10.1016/j.agwat.2016.06.001, 2017.

Kugelman, M.: Pakistan's Interminable Energy Crisis: Is There Any Way Out?, The Wilson Centre, Washington, D.C., https://www.wilsoncenter.org/publication/pakistans-interminable-energy-crisis-there-any-way-out (last access: 1 September 2017), 2015.

Laghari, A. N., Vanham, D., and Rauch, W.: The Indus basin in the framework of current and future water resources management, Hydrol. Earth Syst. Sci., 16, 1063–1083, https://doi.org/10.5194/hess-16-1063-2012, 2012.

Middleton, C., Allouche, J., Gyawali, D., and Allen, S.: The Rise and Implications of the Water–Energy–Food Nexus in Southeast Asia through an Environmental Justice Lens, Water Altern., 8, 627–654, 2015.

Private Power and Infrastructure Board: Hydro power resources of Pakistan, http://www.ppib.gov.pk/HYDRO.pdf (last access: 8 September 2017), 2011.

Roe, E.: Taking complexity seriously: policy analysis, triangulation and sustainable development, Springer, Boston, 2012.

Scott, C. A., Vicuña, S., Blanco-Gutiérrez, I., Meza, F., and Varela-Ortega, C.: Irrigation efficiency and water-policy implications for river basin resilience, Hydrol. Earth Syst. Sci., 18, 1339–1348, https://doi.org/10.5194/hess-18-1339-2014, 2014.

Scott, C. A., Kurian, M., and Wescoat, J. L.: The Water–Energy–Food Nexus: Enhancing Adaptive Capacity to Complex Global Challenges, in: Governing the Nexus, edited by: Kurian, M. and Ardakanian, R., Springer, Cham, 15–38, 2015.

Sharma, R., Markandya, A., Sachdeva, A., Iskakov, M., Krishnaswamy, V., Nikolov, N., and Pedroso, S.: Central Asia Regional Electricity Export Potential Study: Main Report, World Bank, Washington, D.C., 2004

Siddiqi, A. and Wescoat Jr., J. L.: Energy use in large-scale irrigated agriculture in the Punjab province of Pakistan, Water Int., 38, 571–586, https://doi.org/10.1080/02508060.2013.828671, 2013.

UN: The World Population Prospects: 2015 Revision, United Nations Department of Economic and Social Affairs, New York, 2015.

Proc. IAHS, 376, 15–23, 2018
https://doi.org/10.5194/piahs-376-15-2018

Open Access
PIAHS

Supporting better decisions across the nexus of water, energy and food through earth observation data: case of the Zambezi basin

Fortune Faith Gomo[1], Christopher Macleod[2], John Rowan[1], Jagadeesh Yeluripati[2], and Kairsty Topp[3]

[1]Geography, School of Social Sciences, University of Dundee, Dundee, DD1 4HN, UK
[2]The James Hutton Institute, Aberdeen, UK
[3]Scotland's Rural College (SRUC), Edinburgh, EH9 3JG, UK

Correspondence: Fortune Faith Gomo (ffgomo@dundee.ac.uk)

Received: 8 June 2017 – Accepted: 25 July 2017 – Published: 1 February 2018

Abstract. The water–energy–food (WEF) nexus has been promoted in recent years as an intersectional concept designed to improve planning and regulatory decision-making across the three sectors. The production and consumption of water, energy and food resources are inextricably linked across multiple spatial scales (from the global to the local), but a common feature is competition for land which through different land management practices mediates provisioning ecosystem services. The nexus perspective seeks to understand the interlinkages and use systems-based thinking to frame management options for the present and the future. It aims to highlight advantage and minimise damaging and unsustainable outcomes through informed decisions regarding trade-offs inclusive of economic, ecological and equity considerations. Operationalizing the WEF approach is difficult because of the lack of complete data, knowledge and observability – and the nature of the challenge also depends on the scale of the investigation. Transboundary river basins are particularly challenging because whilst the basin unit defines the hydrological system this is not necessarily coincident with flows of food and energy. There are multiple national jurisdictions and geopolitical relations to consider. Land use changes have a profound influence on hydrological, agricultural, energy provisioning and regulating ecosystem services. Future policy decisions in the water, energy and food sectors could have profound effects, with different demands for land and water resources, intensifying competition for these resources in the future. In this study, we used Google Earth Engine (GEE) to analyse the land cover changes in the Zambezi river basin (1.4 million km^2) from 1992 to 2015 using the European Space Agency annual global land cover dataset. Early results indicate transformative processes are underway with significant shifts from tree cover to cropland, with a 4.6 % loss in tree cover and a 16 % gain in cropland during the study period. The changes were found to be occurring mainly in the eastern (Malawi and Mozambique) and southern (Zimbabwe and southern Zambia) parts of the basin. The area under urban land uses was found to have more than doubled during the study period gearing urban centres increasingly as the foci for resource consumption. These preliminary findings are the first step in understanding the spatial and temporal interlinkages of water, energy and food by providing reliable and consistent evidence spanning the local, regional, national and whole transboundary basin scale.

Published by Copernicus Publications on behalf of the International Association of Hydrological Sciences.

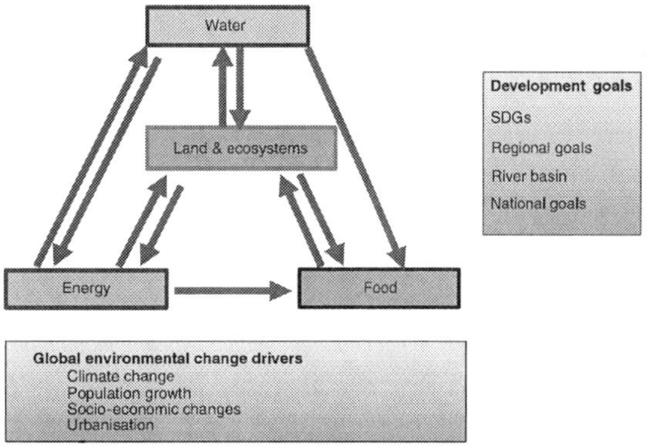

Figure 1. Illustration of linkages between WEF and Land and Ecosystems.

1 Introduction

Land-use and land-cover (LULC) is driven by natural and anthropogenic activities, which in turn drive changes impacting natural ecosystem (Rawat and Kumar, 2015). Understanding the patterns of land cover change and the factors driving spatial and temporal patterns is vital for proper land management and decision use of land for water resources management. There is increasing global stress on land and water resources because of population growth and increased per-capita water, energy and food demand (Pfister et al., 2011). In response, the Water–Energy–Food (WEF) Nexus approach is promoted as a way of understanding the interlinkages between water, energy and food systems. For example, the Bonn 2011 Conference, "The Water Energy and Food Security Nexus – Solutions for the Green Economy" recommended that water, energy and food be considered in an integrative manner, explicitly identifying the interdependencies in decision making. This is based on the premise that decisions made in one sector affect one or more of the other sectors. Integrated decision making is therefore necessary to utilise the synergies and minimise trade-offs. In this study, it is argued water, energy and food rely on the same resource base which is land and ecosystems, and that changes in land cover affect the delivery of water, energy and food ecosystem services. Therefore, land and ecosystems is viewed to be at the centre of the WEF nexus (Fig. 1). The arrows represent the flow of resources from one sector to the other, with land and ecosystems at the centre and supporting all three sectors. Environmental change drivers, such as climate change and population growth are seen to influence decision making to achieve development goals, including Sustainable Development Goals (SDGs), regional, basin and national goals. Table 1 outlines some of the water energy and food ecosystem services associated with major land cover types.

Operationalisation of the WEF Nexus in decision making comes with a set of challenges, some of which were outlined by Efteliogluet al. (2016) as data and modelling constraints. To understand key role of land and ecosystems for understanding and supporting WEF decision making, we need new data such as EO data and more integrated modelling (including web based information/knowledge platforms). Efteliogluet al. (2016) argue that remote sensing in the context of WEF nexus can highlight issues within the nexus and be used to identify resource use hotspots for targeted solutions. "Sub-Saharan Africa has experienced long term LULC but only a few and patchy studies exist on long term characterization of LULC" (Wasige et al., 2013). In the Zambezi, land cover land use studies have also been patchy except for the Zambezi Environment Outlook (ZEO) Report (ZAMCOM, SADC, SARDC, 2015), which describes the changes in the Zambezi basin riparian countries from 2000 and 2010. The report said that there were no significant changes in land cover in the basin from 2000 to 2010.

Kamwi et al. (2015) examined the socio-economic drivers for land use and land cover changes and assessed its impacts to rural livelihoods in the Zambezi region of Namibia. They found that the land cover changes fluctuated, with years of encroachment of cropland and grassland by shrub vegetation as well as clearing of forest vegetation for crops. The authors noted that some of the changes were not permanent, hence the fluctuation. They identified the drivers of land use and land cover change as agricultural expansion, population growth, illegal logging, and increased demand for forest products and to a lesser extent government rules/policy.

Petit et al. (2001) quantified the land cover changes in South Eastern Zambia, in Lusitu area which was reported to have experienced rapid land-cover changes since the resettlement of 6000 people in 1958 because of Kariba dam construction and subsequent filling in 1958. The population of the area rose from around 750 inhabitants pre-Kariba dam, to 8096 in 1963 and by 1994 it had doubled. Since 1958 agriculture has rapidly expanded, increasing bare land situation and increased land degradation. Livestock population also increased from 562 in 1959 to 26 800 in 1991, leading to situations of overgrazing and loss of livestock in times of drought. The authors found that 44 % of the area has been affected by land cover changes with trends towards increasing bare soil and cultivated areas, and rapid decrease in forest and natural vegetation cover. They recorded an average annual rate of change of 4 % with agricultural expansion as the dominant change process. With most of the basin countries having policies geared towards water resources, energy and agricultural development (World Bank, 2010), it is important to understand historical land cover change and land use patterns, to understand spatial interactions. Thus, spatially explicit assessment of water energy and food system/sector interactions is required to identify extent and locations of change (Prestele et al., 2016). This paper is a first step to understand long term land cover change trends in the Zam-

Table 1. Water–energy–food ecosystem services associated with land cover types.

Land Cover/Land Use	Water	Energy	Food
Cropping and herbaceous cover	Water Demand Green water demand Irrigation water Livestock water use	Biofuels production	Food production
Forests	Flow regulation Groundwater recharge Climate regulation	Fuel wood (firewood and charcoal) Carbon sequestration	Game
Shrub land, grassland and sparse vegetation	Flow regulation Climate regulation (lesser extent than forests)		Livestock production Nutrient circulation in soil
Urban/built up areas	Urban water supply Industrial water demand Waste water production	Domestic energy consumption Industrial energy consumption	Urban food demand/consumption (urban farming)
Wetlands and Water bodies	Hydrological regulation/ Flood control Surface water supply Groundwater recharge Water demand Environmental flow requirements Public water use	Hydropower production Cooling water	Flood recession agriculture Irrigation water Livestock water use Fisheries

bezi river basin. The objectives of the study are to assess the changes in land cover in the Zambezi river basin from 1992 to 2015, and to determine the extent of change in cropping systems, forests, urban areas and water bodies. There is a general lack of consistent coverage of satellite data for such large areas, and unavailability of training and validation data for creating land cover maps using raw data, as well as a need for high performance computing (Giri et al., 2013). As a result, there are difficulties in preparing image mosaics and making sure that the multitemporal land cover maps produced for comparison are appropriate for the task (Giri et al., 2013; Vittek et al., 2014). The use of global land cover products allows for the monitoring and analysis of land cover globally and over large areas. It removes the need to work with and process raw satellite images, which would involve volumes of remotely sensed data for such large areas, and deals with problems of data gaps in satellite imagery (Maynard et al., 2016). It is acknowledged here that such data may have limitations, particularly that land cover classification at high resolution may provide limited insight into real land cover dynamics at local scales, and that it can provide preliminary identification of historical and ongoing changes (Tropek et al., 2014).

2 Study area

The study is carried out for the Zambezi River basin. The Zambezi basin covers an area of 1.39 million km^2, the fourth largest in Africa and is one of the most diverse and valuable natural resources in Africa (World Bank, 2010). The basin is transboundary, spanning over 8 countries in Southern Africa, namely Angola, Botswana, Malawi, Mozambique, Namibia, Tanzania, Zambia and Zimbabwe. The basin is home to an estimated population of 32 million people. Figure 2 shows the map of the Zambezi, its sub basins and the riparian countries.

3 Materials and methods

The Climate Change Initiative (CCI) team of the European Space Agency (ESA) officially released the first time series of annual global land cover maps at 300 m resolution, spanning a 24-year period, from 1992 to 2015 in April 2017 (ESA, 2017). The global land cover dataset covers the longest period available to date (ESA, 2017). This high-resolution imagery is considered sufficient to identify areas of land cover change, and estimate changes in the extent (area) for each land cover type in the Zambezi river basin over the 24-year period. More detailed and finer resolution land cover monitoring studies can then be carried out for identified areas of interest. Land cover changes was analysed for the period from 1992 to 2015 to record and assess the major land cover trends in the basin, using Google Earth Engine and ArcGIS 10.5. The ESA/CCI global land cover products used in this study were produced by reprocessing of the full archives of 5 different satellite missions providing daily observation of the Earth, including NOAA-AVHRR HRPT, SPOTVegetation, ENVISAT-MERIS FR and RR, ENVISAT-ASAR, and PROBA-V for the most recent years (ESA, 2017). The land cover changes in the Zambezi

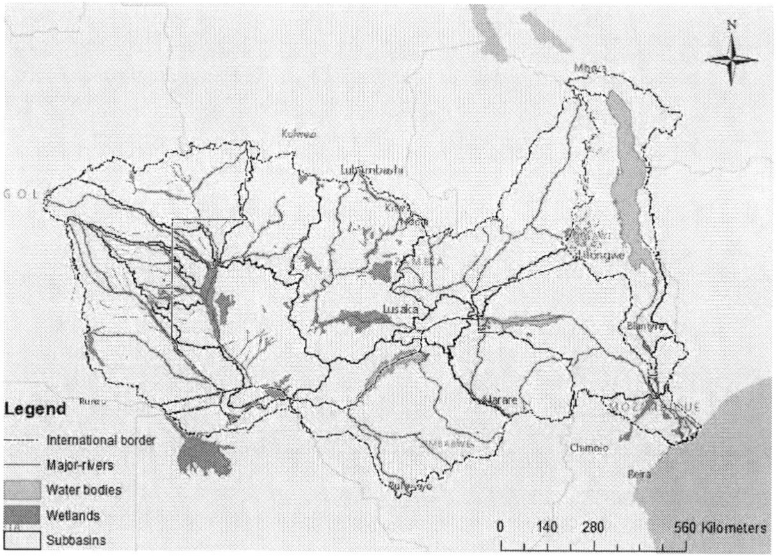

Figure 2. Zambezi Basin Map.

river basin are analysed using Google Earth Engine (GEE) and ArcGIS 10.5. GEE is a web based catalogue of satellite imagery and geospatial datasets with planetary-scale analysis capabilities (https://earthengine.google.com/).

To create a land cover change map, the Zambezi land cover map of 1992 was subtracted from that of 2015. This produced a change vector map with pixels having a spectral reflectance value of the difference between 2015 and 1992. This map showed areas of changed, but even areas of the slightest change showed up on the difference map. The result was squared, and the limit of the spectral variation set at 1000. This meant small differences in the change vector showed up as no change (black) on the map, and those with a squared variation of above 1000 showed up as areas of change (white). In this study, areas of land cover change were defined as those with the square of surface reflectance variation of more than 1000.

Arc GIS 10.5 was then used to quantify the changes in land cover in the basin. Out of the 22 predefined ESA/CCI land cover classes 20 were represented in the Zambezi basin. The Zambezi basin land cover was re-classified into 7 broad classes to represent cropland and herbaceous cover; tree cover; mosaic cropland and natural vegetation; shrub, grassland and sparse vegetation; bare areas; urban areas; and wetlands and water bodies. The areas covered by each of the classes was then calculated from 1992 to 2015. Percentage changes for each broad class were calculated from 1993 to 2015, as follows:

$$\%\,\text{change} = \left(\frac{\text{Area}\,t_1 - \text{Area}\,t_2}{\text{Area}\,t_1}\right) \times 100 \qquad (1)$$

Trend analysis was also carried out for the 7 broad classes to see when the land cover changes occurred and the pattern of change over the study period.

4 Results and discussion

4.1 Land cover changes

There has been progressive clearing of forest areas (green), and expansion of agricultural areas (yellow) since 1992 through 2004 to 2015. This can be seen by simple visual inspection and comparison of the Zambezi land cover maps shown in Fig. 4a and c. These observations are consistent with the findings of Kamwi et al. (2015) who found agricultural activities as the main driver of land cover change in Namibia, and Petit et al. (2001) who found that there were increasing cultivated areas and bare areas in south eastern Zambia from 1986 to 1997. Figure 4d shows the squared change vector map for the Zambezi basin. The black areas show no change while white represents areas where the land cover changed from 1992 to 2015. Figure 4d shows that most noticeable changes occurred in the eastern and southern parts of the basin, with some localized changes in the centre of the basin. Referring to Fig. 4, in the East is Malawi and Mozambique, and in the south, is Zimbabwe and part of Zambia. In the centre of the basin (Zambia) lies some wetlands, and these are also shown to be areas of significant changes in the period 1992 to 2015.

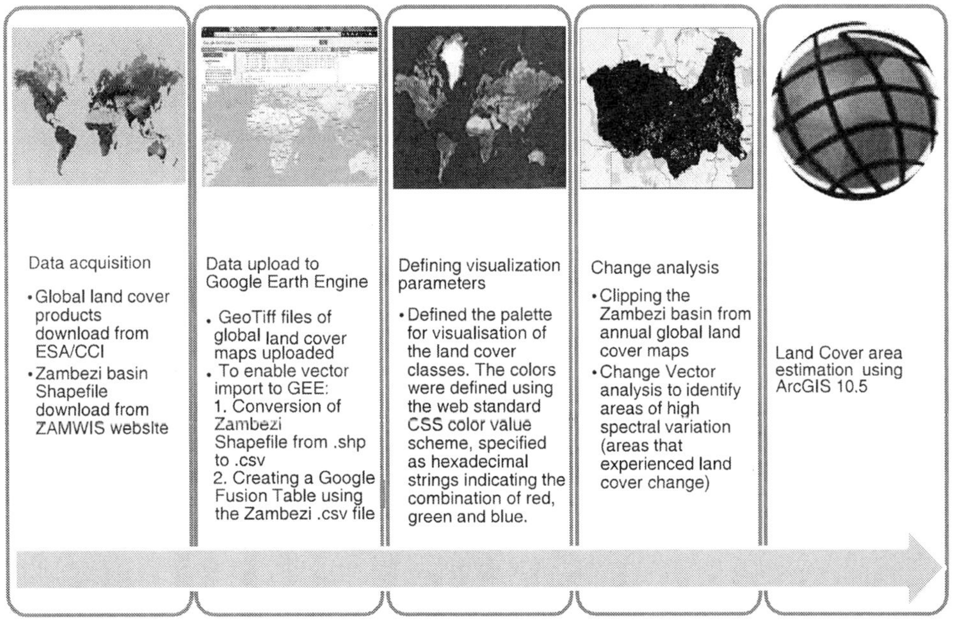

Figure 3. Methodological steps.

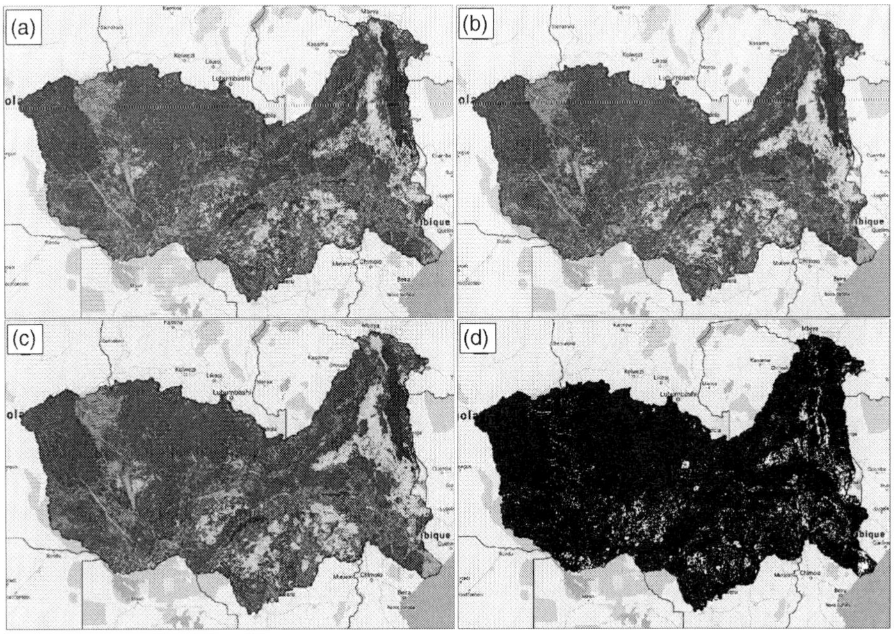

Figure 4. Zambezi land cover maps for 1992 and 2015, and land cover difference map showing areas of land cover change for this period (**a** Zambezi Land Cover 1992; **b** Zambezi land cover 2004; **c** Zambezi land cover 2015; **d** Zambezi squared difference (2015–1992)).

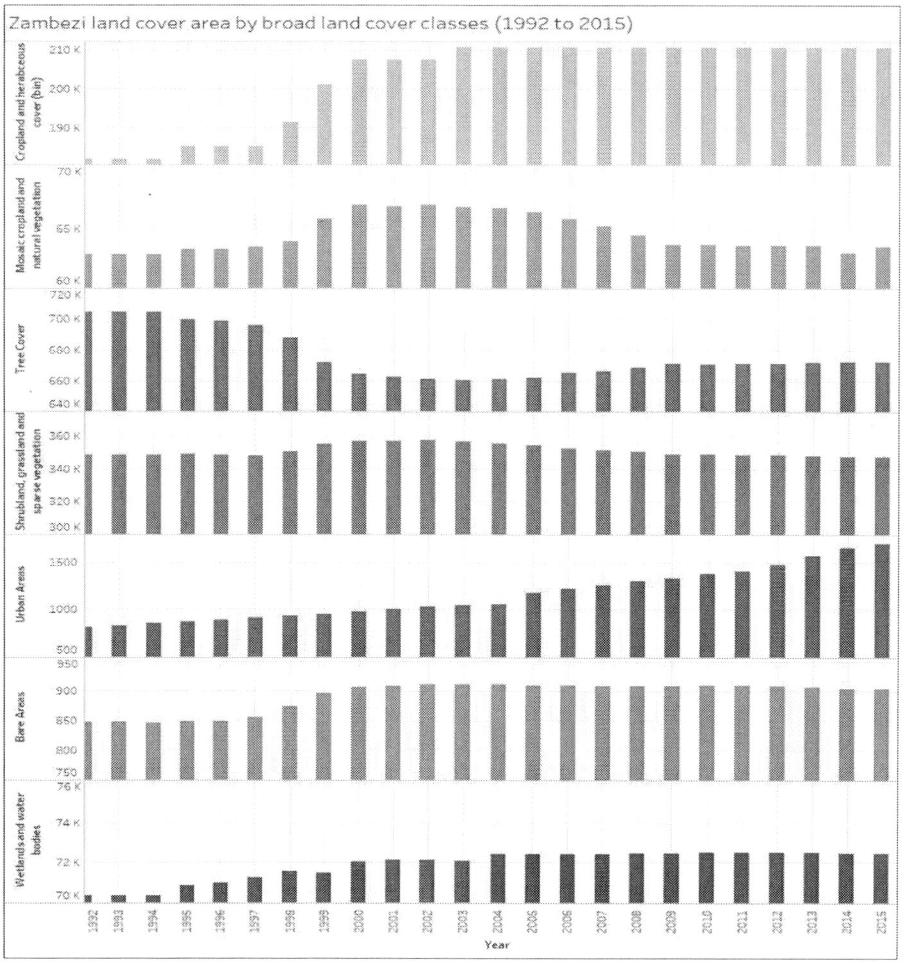

Figure 5. Changes in land cover area (in square kilometres) in the Zambezi river basin.

4.2 Land cover change extent

Having reclassified the land cover into 7 broad classes, the area for each broad class was calculated. Figure 5 shows that the most noticeable changes occurred in forest cover and cropland areas, showing a transformation of tree cover to cropland. Table 2 summarises the percentage changes and shows that cropland area in the Zambezi grew by 15.98 % and covered 15.43 % of the basin area in 2015, while tree cover was lost by 4.57 % between 1992 and 2015. Mosaic of tree cover and cropland increased slightly by 1.1 % and bare areas by 6.78 %, while urban areas more than doubled during the 24 years of study. As much as 3057 km^2 lost for tree cover was not taken up by cropland. This land area may have been taken up by other land cover types that gained, i.e. urban areas, bare areas, mosaics of cropland and natural vegetation. This indicates that trees were not only cleared to make way for cropland, but other land uses too, that may have a bearing on the provision of water, energy and food.

4.3 Land cover change trends

Figure 5 shows the land cover area for the broad classes each year from 1992 to 2015. There was a near exponential increase in cropland from the year 1993 to 2002/3 from at which time the cropland area did not change until 2015.

The pattern of increase in cropland area matches that of the decrease in tree cover area, suggesting that forest areas were cleared to make way for cropland. A similar pattern is noticeable with the mosaic of cropland and natural vegetation, suggesting that the cover transitioned from tree cover to mixed cropland and natural vegetation to cropland. From the year 2006, the mosaic of cropland and natural vegetation area starts to decrease, and this is matched with an in-

Table 2. Percentage changes in land cover from 1992 to 2015.

Land Cover Type	% change 1992–2015	% of basin area in 2015
Cropland and herbaceous cover	15.98 %	15.43 %
Tree Cover	−4.57 %	49.07 %
Mosaic cropland and natural vegetation	1.10 %	4.64 %
Shrubland, grassland and sparse vegetation	−0.19 %	25.38 %
Bare areas	6.78 %	0.07 %
Urban areas	108.66 %	0.12 %
Wetlands and water Bodies	2.95 %	5.29 %

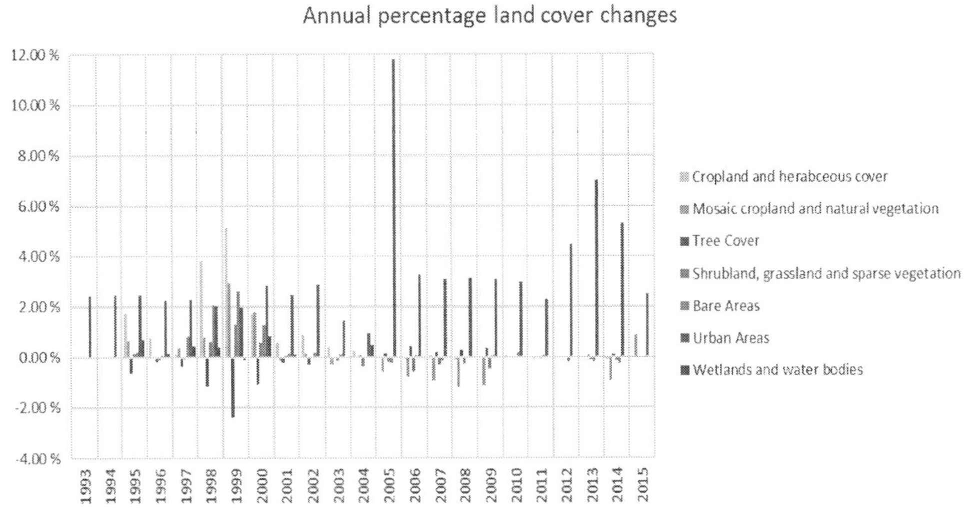

Figure 6. Annual % land cover changes in the Zambezi for the 7 broad classes.

crease in tree cover in the same year, and for the years after, suggesting a regeneration of some of the tree cover that had been cleared earlier for cropland. Urban areas show continued growth throughout the study period, indicating urban expansion in the basin

during the study period. Wetlands and water bodies increased their cover in the early years of the study period, and stayed almost the same from the year 2000 to 2015. Figure 6 shows annual % land cover changes in the Zambezi for the 7 broad classes. Cropland area had the highest annual increase in 1999 (5.17 %), after which the annual changes were below 2 %, and negative in some years up to 2015. In the same year that cropland had the highest increase, tree cover had the highest decrease (2.39 %). Urban areas, expanded by more than double during the study period. There has been a year on year increase of above 2 % in most years of urban areas in the basin, with 50 the highest increase was in 2005 at 11.80 %.

5 Conclusion and discussion

During the period from 1992 to 2015, land cover changes in the Zambezi basin have been mainly occurring in the eastern (Malawi and Mozambique) and southern (Zimbabwe and Zambia) parts of the basin, with some localized changes in the centre (Zambia) of the basin. The localised changes in the centre of the basin were determined to be occurring in wetland areas, where there is loss of wetland vegetation cover.

In the 24-year period of study, there was a gain in cropland area by 15.98 % and a loss of tree cover of 4.57 %. The land area lost in forest cover almost equals the land area gained in cropland, with a difference of 3057 km^2 in land area lost and gained in tree cover and cropland, showing that there is transformation of tree cover to cropland in the basin. This difference could be accounted for clearing of forests for wood fuel as biomass derived energy is the biggest source of energy and accounts for 74 % of energy in the Zambezi basin, and used by most the Zambezi basin population (ZAMCOM, SADC, SARDC, 2015). Land cover change trend analysis showed that cropland increased and the mosaic of cropland

and natural vegetation increased in a pattern similar to that of the decrease of tree cover in the first half of the study period. This shows that there was an expansion of agricultural area during this period, and that there has been no agricultural expansion in the basin during the second half of the study period. This finding is in line with the Zambezi Environment Outlook of 2015, which reported no major changes in cropland from 2000 to 2010. On the other hand, urban areas were found to be the most dynamic land cover type in the basin, increasing steadily throughout the study period, and increasing by more than 100 % since 1992. Identifying and understanding these changes to the basin is vital for informing policy makers, managers and stakeholders about what actions can be taken to better manage land resources to maintain the ecosystem services supporting the WEF nexus effectively. It is important to note that these data are aggregated over a large river basin. The changes recorded here are happening in specific and more localised areas, and it may be worth studying these changes at localised scales at a finer resolution to understand more of the land cover land use dynamics occurring at in these areas.

6 Further work

This paper is a first step to the study of the water, energy and food nexus challenges in the Zambezi river basin. The overall study aims to support decision making in water, energy and food sectors in the basin, by illuminating and clarifying the interlinkages between these sectors, as well as land resources. It is envisaged that this will enable policy and decision makers in the basin to take advantage of the synergies that may be revealed through this study, and minimise tradeoffs among these sectors, thereby making more sustainable decisions to achieve sustainable development goals for water, energy and food. As the Zambezi river basin is vast, a detailed study of this nature for the whole basin would require a bigger research team and more resources. So, as a starting point, this paper identifies resource use hotspots to facilitate a choice of study area within the basin for a more detailed study of the WEF nexus, and how this approach can improve decision making across the WEF sectors.

Code availability. Link to JavaScript code used in GEE for this study: https://code.earthengine.google.com/98a21319dc5b6b22a2c247b4f099ce20. One needs a Google account and Google Earth Engine account to access.

Data availability. Link to ESA Global Land Cover products: http://maps.elie.ucl.ac.be/CCI/viewer/.

Author contributions. This paper is part of a PhD study. FFG is the PhD student and therefore the first author. CJAM, JSR, JY, and CFET are the PhD supervisory team and provide technical and editorial support.

Competing interests. The authors declare that they have no conflict of interest.

Disclaimer. The views and opinions expressed in this article are those of the authors and do not necessarily reflect the official policy or position of project funders, the Scottish Government and the HydroNation Programme. This article is a product of academic research.

Special issue statement. This article is part of the special issue "Water security and the food–water–energy nexus: drivers, responses and feedbacks at local to global scales". It is a result of the IAHS Scientific Assembly 2017, Port Elizabeth, South Africa, 10–14 July 2017.

Acknowledgements. This paper is part of a PhD project funded by the Scottish Government through HydroNation Scholars Programme 2015 (grant code 064272.S002). The authors would also like to thank Mark Cutler for his helpful discussions.

Edited by: Barry Croke
Reviewed by: Lauren Bulcock and one anonymous referee

References

Eftelioglu, E., Jiang, Z., Ali, R., and Shekhar, S.: Spatial computing perspective on food energy and water nexus, J. Environ. Stud. Sci., 56, 62–76, 2016.

European Space Agency: 300 m annual global land cover time series from 1992 to 2015, Climate Change Initiative Land Cover Newsletter, Issue n. 7 April 2017, Retrieved on 10 April 2017 from https://www.esa-landcover-cci.org/?q=webfm_send/88, last access: 1 September 2017.

Giri, C., Pengrab, B., Longc, J., and Lovelanda, T. R.: Next generation of global land cover characterization, mapping, and monitoring, Int. J. Appl. Earth Obs., 25, 30–37, https://doi.org/10.1016/j.jag.2013.03.005, 2013.

Kamwi, J. M., Chirwa, P. W. C., Manda, S. O. M., Graz, P. F., and Kätsch, C.: Livelihoods, land use and land cover change in the Zambezi Region, Namibia, Popul. Environ., 37, 207–230, 2015.

Maynard, J. J., Karl, J. W., and Browning, D. M.: Effect of spatial image support in detecting long-term vegetation change from satellite time-series, Landscape Ecol., 31, 2045, https://doi.org/10.1007/s10980-016-0381-y, 2016.

Petit, C., Scudder, T., and Lambin, E.: Quantifying processes of land-cover change by remote sensing: Resettlement and rapid land-cover changes in southeastern Zambia, Int. J. Remote Sens., 22, 3435–3456, https://doi.org/10.1080/01431160010006881, 2001.

Pfister, S., Bayer, P., Koehler, A., and Hellweg, S.: Environmental impacts of water use in global crop production: Hotspots and trade-offs with land use, Environmental Science and Technology, 45, 5761–5768, 2011.

Prestele, R., Alexander, P., Rounsevell, M. D. A., Arneth, A., Calvin, K., Doelman, J., Eitelberg, D. A., Engström, K., Fujimori, S., Hasegawa, T., Havlik, P., Humpenöder, F., Jain, A. K., Krisztin, T., Kyle, P., Meiyappan, P., Popp, A., Sands, R. D., Schaldach, R., Schüngel, J., Stehfest, E., Tabeau, A., Van Meijl, H., Van Vliet, J., and Verburg, P. H.: Hotspots of uncertainty in land-use and landcover change projections: a global-scale model comparison, Glob. Change Biol., 22, 3967–3983, 2016.

Rawat, J. S. and Kumar, M.: Monitoring land use/cover change using remote sensing and GIS techniques: A case study of Hawalbagh block, district Almora, Uttarakhand, India, The Egyptian Journal of Remote Sensing and Space Science, 18, 77–84, 2015.

Tropek, R., Sedláček, O., Beck, J., Keil, P., Musilová, Z., Šímová, I., and Storch, D.: Comment on "High-resolution global maps of 21st-century forest cover change", Science, 344, 6187, https://doi.org/10.1126/science.1248753, 2014.

Vittek, M., Brink, A., Donnay, F., Simonetti, D., and Desclée, B.: Land Cover Change Monitoring Using Landsat MSS/TM Satellite Image Data over West Africa between 1975 and 1990, Remote Sensing, 2014, 658–676, https://doi.org/10.3390/rs6010658, 2014.

Wasige, J. E., Groen, T. A., Eric, S., and Victor, J.: Monitoring basin-scale land cover changes in Kagera basin of lake Victoria using ancillary data and remote sensing, Int. J. Appl. Earth Obs., 21, 32–42, https://doi.org/10.1016/j.jag.2012.08.005, 2013.

World Bank: Summary report, Washington, DC, World Bank, http://documents.worldbank.org/curated/en/724861468009989838/Summary-report (last access: 1 September 2017), 2010.

ZAMCOM, SADC, SARDC: Zambezi Environment Outlook 2015, ZAMCOM, SADC,SARDC, Harare, Zimbabwe, Gaborone, Botswana, 2015.

Proc. IAHS, 376, 25–33, 2018
https://doi.org/10.5194/piahs-376-25-2018

Open Access
PIAHS

Water impacts and water-climate goal conflicts of local energy choices – notes from a Swedish perspective

Rebecka Ericsdotter Engström[1], Mark Howells[1], and Georgia Destouni[2]

[1]Department of Energy Technology, Royal Institute of Technology, Stockholm, Sweden
[2]Department of Physical Geography and the Bolin Centre of Climate Research, Stockholm University, Stockholm, Sweden

Correspondence: Rebecka Ericsdotter Engström (rebecka.engstrom@energy.kth.se)

Received: 6 June 2017 – Accepted: 30 August 2017 – Published: 1 February 2018

Abstract. To meet both the Paris Agreement on Climate Change and the UN Sustainable Development Goals (SDGs), nations, sectors, counties and cities need to move towards a sustainable energy system in the next couple of decades. Such energy system transformations will impact water resources to varying extents, depending on the transformation strategy and fuel choices. Sweden is considered to be one of the most advanced countries towards meeting the SDGs. This paper explores the geographical origin of and the current water use associated with the supply of energy in the 21 regional counties of Sweden. These energy-related uses of water represent indirect, but still relevant, impacts for water management and the related SDG on clean water and sanitation (SDG 6). These indirect water impacts are here quantified and compared to reported quantifications of direct local water use, as well as to reported greenhouse gas (GHG) emissions, as one example of other types of environmental impacts of local energy choices in each county. For each county, an accounting model is set up based on data for the local energy use in year 2010, and the specific geographical origins and water use associated with these locally used energy carriers (fuels, heat and electricity) are further estimated and mapped based on data reported in the literature and open databases. Results show that most of the water use associated with the local Swedish energy use occurs outside of Sweden. Counties with large shares of liquid biofuel exhibit the largest associated indirect water use in regions outside of Sweden. This indirect water use for energy supply does not unambiguously correlate with either the local direct water use or the local GHG emissions, although for the latter, there is a tendency towards an inverse relation. Overall, the results imply that actions for mitigation of climate change by local energy choices may significantly affect water resources elsewhere. Swedish counties are thus important examples of localities with large geographic zones of water influence due to their local energy choices, which may compromise water security and the possibility to meet water-related global goals in other world regions.

1 Introduction

Increased emphasis on local action for global change in the sustainability discourse encourages efforts of sub-national authorities to prioritize sustainability (Theobald et al., 2015; Wamsler et al., 2014; Xue and Tobias, 2015, among others). International initiatives to promote local sustainability still tend to focus on local production-based resource uses and emissions (Covenant of Mayors, 2017). However, trade-related environmental impacts are now increasingly analysed through e.g. concepts of water footprint and life-cycle, and tracking of consumption-based (as opposed to more common production-based) greenhouse gas (GHG) emissions (Mekonnen and Hoekstra, 2010; Gerbens-Leenes et al., 2012; Tukker, 2000; Stokes and Horvath, 2010; Caro et al., 2017; Yang et al., 2015). Moreover, in the wake of the adoption of the UN Sustainable Development Goals (SDGs) in 2015, the need for integrated and cross-scale coordination of sustainability actions is increasingly acknowledged (Nilsson et al., 2016, among others).

Two SDGs with particularly strong interactions are those focused on water (SDG 6) and energy (SDG 7). Both resources are crucial for human survival and prosperity, and

Published by Copernicus Publications on behalf of the International Association of Hydrological Sciences.

both need to be managed more sustainably in the coming decades in order for the world to meet the related SDGs (Griggs et al., 2017). Direct (at the energy utility) and indirect (throughout the whole energy supply chain) uses of water for electricity production have been shown to be significant, not only on local but also on, regional continental and global scale (Destouni et al., 2013; Jaramillo and Destouni, 2015). The *water-energy nexus* has also been increasingly studied in the past decade (among many others: Hamiche et al., 2016; Hussey and Pittock, 2012; Bazilian et al., 2011; Scott et al., 2011). The ability to accurately assess both direct-local and indirect-remote interactions of water and energy uses will only increase in importance as local actors move towards implementing the SDGs. The present study addresses this water-energy nexus on different scales.

Specifically, we here quantify and analyse *energy*-related indirect-remote water use, i.e. the water-use required for extracting, processing, transforming and supplying energy for local use, using the 21 counties of Sweden as concrete case examples of such local energy use; Fig. 1 displays the 21 counties on the map of Sweden. In general, Sweden is considered to be the most advanced country towards meeting the SDGs (Sachs et al., 2016). As such, the present case study focus on and across different Swedish counties may provide some important insights on relatively advanced water-impact patterns associated with local energy choices towards long-term sustainability. The analysis is further broadened, to also consider the climate-related SGD (13), by relating the water-energy nexus results to the officially reported GHG emissions of each county.

The used data and obtained results of this study (direct water use, GHG emissions, and calculated energy-related indirect water use) are not overlapping in scope (the latter includes trade impacts, while the former two relate only to local activities). This scope inconsistency is intentional, in order to clarify the difference between data that are readily available for – and directly relate to the mandate of – county-level decision-makers (reported local direct water use and GHG emissions) and the indirect energy-related water impacts (indirect-remote water use for energy supply) that are calculated in this study and are significant for achievement of global sustainability.

In the following sections, the analytical approach and data sources are described, followed by presentation and discussion of results and main conclusions.

2 Method

2.1 Analytical approach

In the present analysis, a simple accounting framework is employed to assess the water impacts of the energy choices of different Swedish counties. The following steps are taken for this purpose:

Figure 1. Map of Sweden with the 21 counties studied.

- Step 1: a literature review of reported water-consumption estimates for energy production including: water footprints of biofuels; consumptive water-use of electricity generation (incl. hydropower) and; water use in fossil-fuel production processes (mining, refining and distribution) (the terms used here correspond to those used in the reviewed literature).
- Step 2: multiplication of water factors, calibrated for Swedish energy-supply specificity after the general Step 1, with disaggregated energy end-use by sector and fuel-type for each of Sweden's 21 counties (accounting also for energy transmission losses – see Supplement for further methodological details).

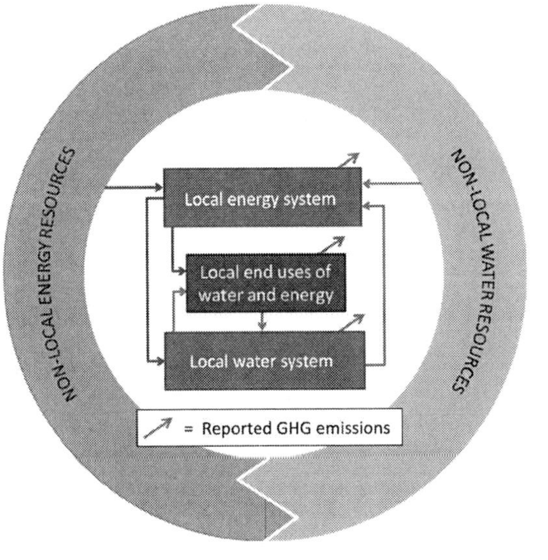

Figure 2. Conceptual CLEWs mapping of the water-energy interactions assessed in this study.

- Step 3: disaggregation of indirect water-use between national and international, in order to assess the energy-related water use occurring outside of Sweden (and as such outside the nation's sustainability-reporting boundaries).
- Step 4: comparison of results with reported county-specific GHG emissions and direct water use.

Only freshwater use is considered in this analysis, in terms of both direct and indirect water uses. Seawater, often used in e.g. cooling systems of nuclear power plants (Vattenfall, 2016), is excluded from the analysis since this water resource is not included in the water-related SDG (6) and also is not subject to similar availability constraints as the freshwater resource. A schematic of the system-of systems considered in this study is depicted in Fig. 2.

2.2 Data sources

Overall, official county-level data for the year 2010 is used in the present analysis, and water-use results for various energy choices are compared with official statistics on county-specific direct water use and GHG emissions for the same year. The GHG emission data are taken from the Swedish County Board Union's yearly report on airborne emissions (Nationella emissionsdatabasen, 2017). Direct water-use data are collected and published every 5th year by the Swedish Central Bureau of Statistics (SCB) and for this paper tables MI0902AB and 0000000V are used (SCB Statistikdatabasen, 2017). Available SCB data (table EN0203AE) are also used for county-specific energy use, disaggregated over fuel types (liquid fossil, solid renewables, electricity, district heat, etc.) and over user groups (residential buildings, industry, public sector, etc.) (SCB Statistikdatabasen, 2017).

Data on water use for energy supply

The water-energy nexus of fuel production and electricity and heat generation has been extensively studied in recent years (e.g. Bakken et al., 2013; EPRI, 2008; Holland et al., 2015; Macknick et al., 2012). Still, reported data vary greatly with geographical and technical conditions for the energy systems assessed, and also due to large variation in methods used to measure or indirectly analyse water use. The need to improve both data availability and data standards in this field has been stressed by, among others, Macknick et al. (2012). Awaiting more such efforts to been made, this study uses available data "at face value", in the attempt to gain insights on patterns and orders of magnitude, while acknowledging the dependence of results on such greatly varying and uncertain data. For the present analysis, a database on published water factors (in units of water volume per energy-unit produced) is created based on the following studies and reports: Mielke et al. (2010); Pacetti et al. (2015); Gerbens-Leenes et al. (2008); Scown et al. (2011); Granit and Lindström (2010); Mekonnen et al. (2015); Fthenakis and Kim (2010); Katers et al. (2012); Rio Carrillo and Frei (2009); Spang et al. (2014); IEA (2012). Considering the geographical origins as well as the types of fuels (conventional or unconventional fossil fuels, first or second generation biofuels, etc.) used in Sweden, the data points that most closely resemble Swedish conditions are selected from this database for the present analysis. Table 1 summarizes the water factors chosen for the analysis.

3 Results

Figures 3 and 4 display the calculated indirect water-use requirements for the energy supply in the 21 counties of Sweden, in absolute and per-capita terms, respectively. The size of circles in these graphs corresponds to the relative volume of energy-related indirect freshwater use, with the x axis representing reported county emissions of GHGs and the y axis representing county-specific direct local water-use.

In terms of *absolute* values (Fig. 3), the most populous counties (Stockholm, Skåne and Västra Götaland) have the largest reported GHG emissions, as well as most of the largest reported direct water uses. However, also the county of Norrbotten (in terms of emissions) and that of Västernorrland (in terms of direct water use) display high absolute resource uses despite having much smaller populations. Västernorrland has the most water-intensive industry in Sweden, while Norrbotten is the county using the most coal of all Swedish counties in 2010, with thereby high associated GHG emission levels. These latter two counties stand out even more clearly in Fig. 4, where results are displayed in terms of per-capita units. In addition, the county with the

Table 1. Water factors employed in the present analysis.

Fuel	Water factor (m^3 TJ^{-1})	Origin of Data Reference/ Origin of Fuel in Swedish Supply	Data Reference
Oil	259	USA/Russia	Mielke et al. (2010)
Gasoline	283	USA/Russia	Scown et al. (2011)
Diesel	274	USA/Russia	Mielke et al. (2010)
Natural Gas	4	NA/Denmark	Spang et al. (2014)
Coal	43	Latin America/USA	Spang et al. (2014)
Biodiesel/Bio-oil	19 800	NA/Germany	Spang et al. (2014)
Ethanol	24 700	Brazil/Brazil	Spang et al. (2014)
Biogas	149	Italy/Sweden	Calculated based on data from Pacetti et al. (2015)
Pellets/biomass	42.6	USA/Sweden	Katers and Snippen (2012)
Electricity	2310	Sweden/Sweden	Calculated based on Swedish electricity mix. Details in the Supplement.
District Heat	795	Sweden/Sweden	Calculated based on Swedish average district heat fuel mix. Details in the Supplement.

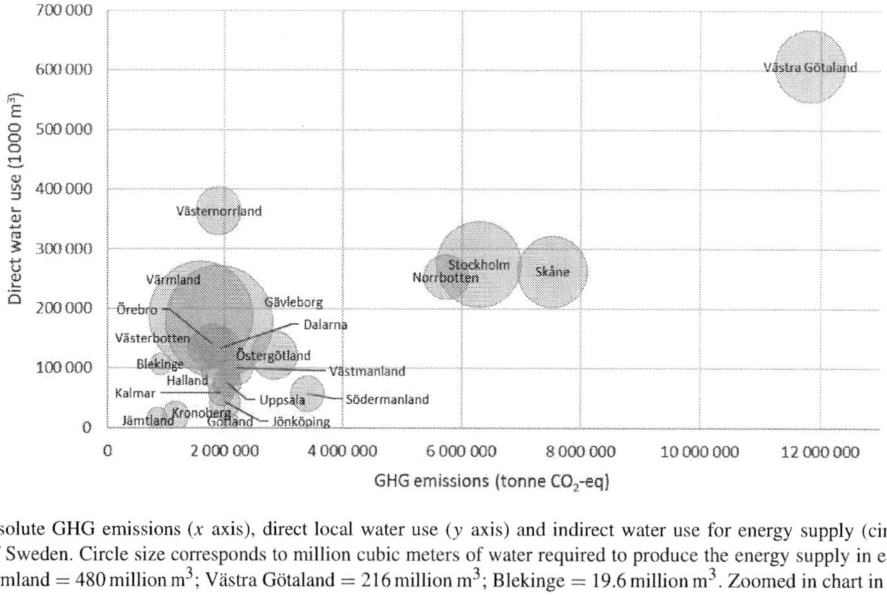

Figure 3. Absolute GHG emissions (x axis), direct local water use (y axis) and indirect water use for energy supply (circle size) in the 21 counties of Sweden. Circle size corresponds to million cubic meters of water required to produce the energy supply in each county. For reference: Värmland = 480 million m^3; Västra Götaland = 216 million m^3; Blekinge = 19.6 million m^3. Zoomed in chart in Supplement.

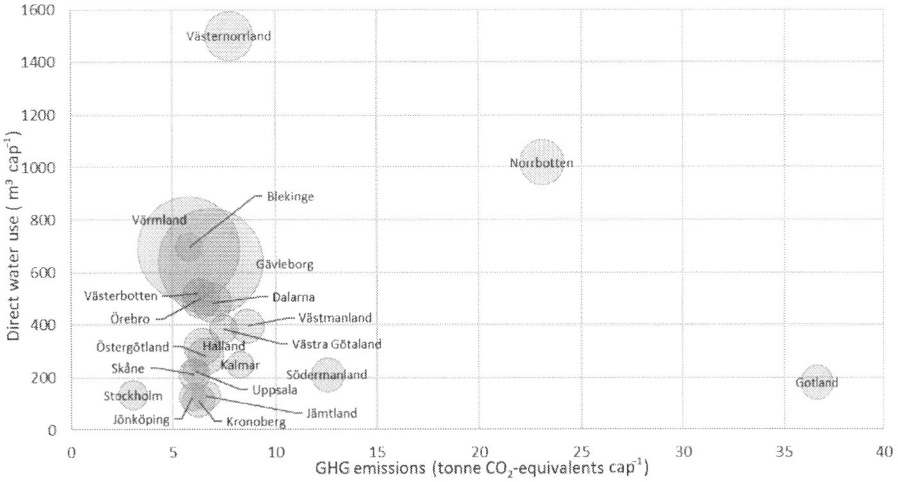

Figure 4. Per-capita GHG emissions (x axis), per-capita direct local water use (y axis) and per-capita indirect water use for energy supply (circle size) in the 21 counties of Sweden. Circle size corresponds to per-capita cubic meters of water indirectly used to produce the energy supply in each county. For reference: Värmland = 1756 $m^3\ cap^{-1}$; Gotland = 200 $m^3\ cap^{-1}$; Blekinge = 128 $m^3\ cap^{-1}$. Zoomed in chart in Supplement.

smallest population, Gotland, stands out as the county with the highest per capita GHG emissions, due to significant fossil GHG emissions from the county's lime and cement industries (Region Gotland, 2016). However, the direct local water use in Gotland is limited, with this being a Swedish county that is particularly prone to water shortages and droughts.

Looking further at the results of indirect water use for energy supply (circle sizes in Figs. 3 and 4), the pattern differs from both that of reported GHG emissions and that of reported direct water use. Most notable here are the counties of Värmland and Gävleborg, with significantly larger energy-related indirect use of water than that of other Swedish counties. The reason is that the industries in Värmland and Gävleborg use large volumes of renewable liquid fuels (assumed to correspond to biodiesel or bio-oil in Table 1). As shown in Table 1, these fuels consume 70–90 times more water per unit of energy than corresponding liquid fossil fuels (and this is a modest water-use estimate compared to those of Gerbens-Leenes et al., 2009; IEA, 2012; among others).

Gotland's indirect per-capita use of water does not follow the pattern of its per-capita emissions of GHGs. As mentioned, this county has a relatively high share of fossil GHG emissions but uses almost no liquid biofuels. This choice makes the county's energy use relatively water efficient.

In Fig. 5, the national (Swedish) indirect water use (for fuel, electricity and heat production within Sweden) is removed from the results. This exacerbates the pattern seen in Fig. 4, exhibiting Gävleborg and Värmland as responsible for an even larger share of Sweden's indirect (imported) energy-related water use (Fig. 5). This is not surprising given the large water requirements of imported liquid biofuels (in the present study, no domestic liquid biofuels are considered available). It also emphasizes the large discrepancy that may prevail between the direct water use in local service-delivery systems and the non-local and indirect water use required for the local energy supply.

If direct water use were a good indicator of total water use (including indirect water from energy imports), the size of the circles in Figs. 3–5 should generally increase along the y axis, which is not the case. Similarly, and even though both water use and GHG emissions are tightly linked to the energy sector, the levels of GHG emissions are not correlated with the remote and indirect water use for the local energy supply (as circle sizes do not systematically increase (or decrease) along the x axis).

For further analysis of the joint water-use and GHG connections to energy, Fig. 6 displays all discussed water-use and emission results, as well as the total energy use in each county. Värmland and Västernorrland are then found to have similar profiles of per-capita energy use and per-capita GHG emissions (boxed part of Fig. 6). However, their indirect per-capita water uses for energy are far apart. In Värmland, the chosen low-carbon energy use consists to large degree of liquid biofuels, while the per-capita energy use of Västernorrland includes primarily electricity and biomass – two fuels that consume much less water than imported liquid biofuels. These different choices explain some of the discrepancy seen in Fig. 6 for indirect water-use per capita. However, in this context, it is again worth noting that the used data on energy-related GHG emissions and the calculated data on indirect water use have different scopes of measure. The former (GHG data) include local emission sources, including burn-

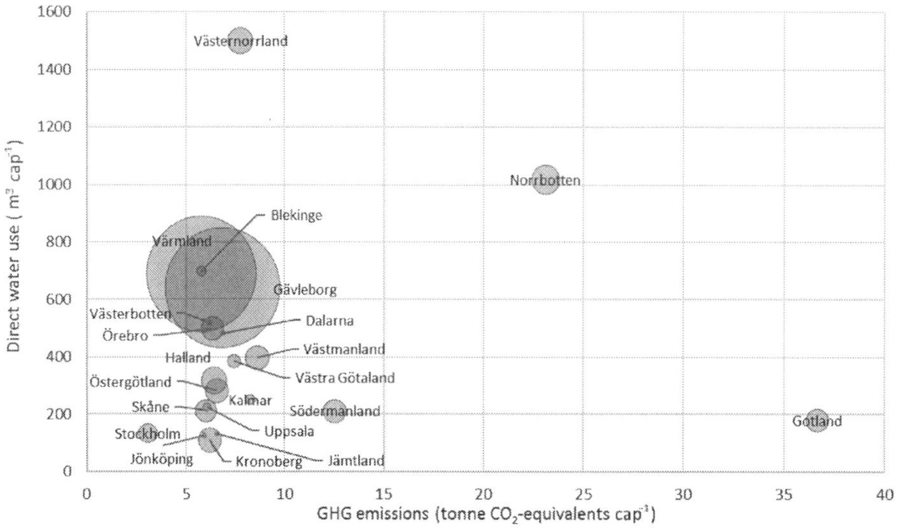

Figure 5. "Imported" indirect water use outside of Sweden related to the population (per-capita units) and the energy use in the 21 counties of Sweden, compared to reported direct water use per capita (y axis) and GHG emissions per capita (x axis). Circle size corresponds to per capita cubic meters of water indirectly used outside the national borders to produce the energy supply in each county. Note: For reference: Värmland = 1581 m^3 cap^{-1}; Gotland = 62 m^3 cap^{-1}; Blekinge = 13 m^3 cap^{-1}. Zoomed in chart in Supplement.

ing of fuels in vehicles and in electricity- and heat-production utilities, etc., locally as well as local non-combustion/energy related emissions from agriculture, waste facilities and similar. The latter (data on water use for energy supply) consider only the energy sector and the energy *use* in each county, including the full energy supply chain, both local and non-local. Although we can largely explain the patterns of GHG emissions and of indirect water use by investigating the specific mix of fuels used in a county, there is no unambiguous pattern of e.g. low (high) emissions corresponding to high (low) indirect water use. With that said, if Värmland's industries had chosen to use fossil fuels instead of biofuels, one would expect higher GHG emission levels in the county and lower indirect water use associated with the county's energy supply.

Finally, looking at the combination of direct water use and GHG emissions reported, Figs. 4 and 5 show that Blekinge has similar x/y-coordinates (per-capita direct water use and emissions, respectively) as Värmland. However, their indirect water uses for energy supply (circle size) are very different. As such, neither direct water use nor GHG emissions, or the combination of these, provides any clear clue to the indirect water use of a county. Specific case-by-case analysis of the considered energy fuels – and ideally a more focused analysis of the specific origin of these fuels – is needed to assess this type of externalized water impacts with greater certainty. Nevertheless, the present analysis supports existing work on water-energy resource interdependencies (Hussey and Pittock, 2012, among others) and the suggested importance of SDG interactions by Nilsson et al. (2016). The studied external (remote) water impacts of local energy strategies are significant, but difficult to trace with great precision and certainty from only local-level resource-use and emissions data.

As noted in the data section, limitations in both availability and comparability of data on water use in energy supply chains are significant. Although careful selection of water factors has been made for this analysis (Table 1), the uncertainties and temporal and spatial variations of reported energy-related water uses are large. As a consequence, the results presented here should be interpreted and used conservatively. This work primarily aims to highlight the un-intuitive (sometimes counter-intuitive) relations between direct and indirect environmental impacts of local energy choices, and the potentially large differences in indirect resource impacts (in our study focused on water) of regions with seemingly comparable local emissions profiles and direct resource uses.

4 Conclusions

Focusing on the water-energy nexus, this paper has identified quite different patterns of reported direct water use, GHG emissions and indirect water use for local energy use among Sweden's 21 counties. Direct water use is typically driven by population size and the presence of water-intensive industries. In contrast, energy-related indirect water use is primarily impacted by the choice of fuels used in the county – with liquid biofuels requiring orders of magnitude larger volumes

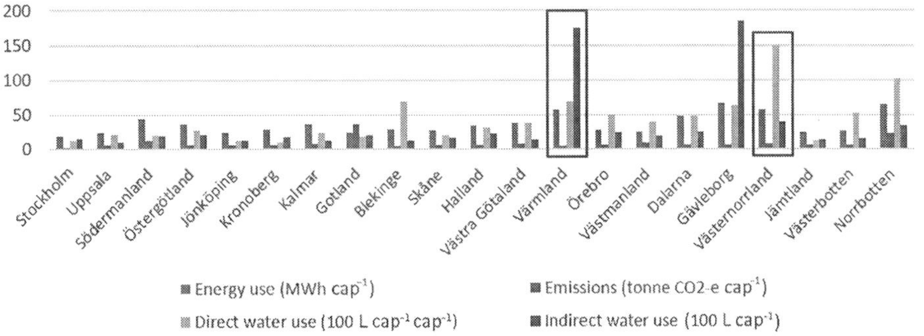

Figure 6. Summary of per-capita county results for: direct energy use; GHG emissions; direct water use; and energy-related indirect water use. Red boxes highlights 2 counties with very similar energy use and emissions profiles, but large differences in both direct water use and indirect – energy related – water use.

of water than all other fuels used in the analysed energy systems.

Similar to the indirect water use studied in this paper, GHG emissions are tightly coupled to the energy sector. However, on a fuel-by-fuel basis, the water-use and GHG interactions with energy differ significantly. Conventional fossil fuels, with their high CO_2 emissions, are orders of magnitude less water-intensive than the carbon-neutral biofuels.

As climate-change action plans are developed and adopted by an increasing number of sub-national authorities (to meet the Paris Agreement as well as the climate-related SDG 13), the goal conflict implied by a largely (even though not fully) inverse relation between GHG emissions and indirect water use should be acknowledged and further investigated. As no clear general relation could be determined in the present analysis of energy-related GHG emissions, indirect-remote water use and direct-local water use, further investigation of such relations are needed for different local-regional cases. Such further analysis is not least needed in Sweden, if the country is to remain a front-runner in reaching the highly integrated SDGs, with the freshwater SGD (6), the energy SDG (7) and the climate SDG (13) considered as equally important sustainability targets.

Finally, accounting for this type of cross-scale interactions in the monitoring of progress towards reaching the SDGs could potentially alter assessments of how different nation states or sub-national regions are performing, and should be considered when indicators for monitoring progress towards the SDGs are further developed.

Data availability. Data sources are cited throughout the text and come from:

- water use for energy: published literature (as referenced in Table 1 and Table S1)
- energy use and water use per county: the Swedish Central Bureau of Statistics, SCB (Statistiska Centralbyrån), data tables: EN0203AE (energy use), MI0902AB (water use) and 0000000V (water use). Data for year 2010
- GHG emissions per county: RUS, Nationella Emissionsdatabasen. Data for year 2010

The Supplement related to this article is available online at https://doi.org/10.5194/piahs-376-25-2018-supplement.

Competing interests. The authors declare that they have no conflict of interest.

Special issue statement. This article is part of the special issue "Water security and the food–water–energy nexus: drivers, responses and feedbacks at local to global scales". It is a result of the IAHS Scientific Assembly 2017, Port Elizabeth, South Africa, 10–14 July 2017.

Edited by: Barry Croke
Reviewed by: James Cullis and two anonymous referees

References

Bakken, T. H., Killingtveit, Å., Engeland, K., Alfredsen, K., and Harby, A.: Water consumption from hydropower plants – review of published estimates and an assessment of the concept, Hydrol. Earth Syst. Sci., 17, 3983–4000, https://doi.org/10.5194/hess-17-3983-2013, 2013.

Bazilian, M., Rogner, H., Howells, M., Hermann, S., Arent, D., Gielen, D., Steduto, P., Mueller, A., Komor, P., Tol, R., and Yumkella, K.: Considering the Energy, Water and Food Nexus: Towards an Integrated Modelling Approach, Energ. Policy, 39, 7896–7906, https://doi.org/10.1016/j.enpol.2011.09.039, 2011.

Caro, D., Pulselli, F. M., Borghesi, S., and Bastianoni, S.: Mapping the International Flows of GHG Emissions within a More

Feasible Consumption-Based Framework, J. Clean. Prod., 147, 142–151, https://doi.org/10.1016/j.jclepro.2017.01.106, 2017.
Covenant of Mayors: http://www.covenantofmayors.eu/index_en.html, last access: 28 March 2017.
Destouni, G., Jaramillo, F., and Prieto, C.: Hydroclimatic Shifts Driven by Human Water Use for Food and Energy Production, Nature Climate Change, 3, 213–217, https://doi.org/10.1038/nclimate1719, 2013.
EPRI: Water Use for Electric Generation, Palo Alto, CA, 1014026, available at: https://www.epri.com/#/pages/product/1014026/ (last access: 29 May 2017), 2008.
Fthenakis, V. and Kim, H. C.: Life-Cycle Uses of Water in U.S. Electricity Generation, Renewable and Sustainable Energy Reviews, 14, 2039–2048, https://doi.org/10.1016/j.rser.2010.03.008, 2010.
Gerbens-Leenes, P. W., Hoekstra, A. Y., and van der Meer, Th. H.: Water Footprint of Bio-Energy and Other Primary Energy Carriers, Value of water research report series 29. UNESCO-IHE Institute for Water Education, available at: https://ris.utwente.nl/ws/portalfiles/portal/5148108 (last access: 10 October 2017), 2008.
Gerbens-Leenes, P. W., Hoekstra, A. Y., and van der Meer, Th. H.: The Water Footprint of Bioenergy, Proceedings of the National Academy of Sciences, 106, 10219–10223, https://doi.org/10.1073/pnas.0812619106, 2009.
Gerbens-Leenes, P. W., Van Lienden, A. R., Hoekstra, A. Y., and Van der Meer, Th. H.: Biofuel Scenarios in a Water Perspective: The Global Blue and Green Water Footprint of Road Transport in 2030, Global Environmental Change, 22, 764–775, https://doi.org/10.1016/j.gloenvcha.2012.04.001, 2012.
Granit, J. and Lindström, A.: Constraints and Opportunities in Meeting the Increasing Use of Water for Energy Production, in: Proceedings of the ESF Strategic Workshop on Accounting for Water Scarcity and Pollution in the Rules of International Trade, Value of Water Research Report Series. Amsterdam: UNESCO-IHE Institute for Water Education, 54, 85–98, 2010.
Griggs, D., Nilsson, M., Stevance, A., and McCollum, D.: A Guide to SDG Interactions: From Science to Implementation, Report, Paris, France: International Council for Science (ICSU), http://www.icsu.org/cms/2017/05/SDGs-Guide-to-Interactions.pdf, 2017.
Hamiche, A. M., Boudghene Stambouli, A., and Flazi, S.: A Review of the Water-Energy Nexus, Renew. Sust. Energ. Rev., 65, 319–331, https://doi.org/10.1016/j.rser.2016.07.020, 2016.
Holland, R. A., Scott, K. A., Flörke, M., Brown, G., Ewers, R. M., Farmer, E., Kapos, V., Muggeridge, A., Scharlemann, J. P. W., Taylor, G., Barrett, J., and Eigenbrod, F.: Global Impacts of Energy Demand on the Freshwater Resources of Nations, P. Natl. Acad. Sci., 112, E6707–E6716, https://doi.org/10.1073/pnas.1507701112, 2015.
Hussey, K. and Pittock, J.: The Energy–Water Nexus: Managing the Links between Energy and Water for a Sustainable Future, Ecol. Soc., 17, 31, https://doi.org/10.5751/ES-04641-170131, 2012.
IEA: World Energy Outlook 2012, Paris, OECD Publishing, https://doi.org/10.1787/weo-2012-en, 2012.
Jaramillo, F. and Destouni, G.: Local Flow Regulation and Irrigation Raise Global Human Water Consumption and Footprint, Science, 350, 1248–1251, https://doi.org/10.1126/science.aad1010, 2015.
Katers, J. F. and Snippen, A.: Life-cycle Inventory of Wood Pellets Manufacturing in Wisconsin, Final Report, Environmental and Economic Research and Development Program of the Public Service Commission of Wisconsin & the Statewide Energy Efficiency and Renewables Administration, available at: https://www.focusonenergy.com/sites/default/files/research/katerswoodpelletmfg_report.pdf, (last access: 30 September 2017), March 2011.
Macknick, J., Sattler, S., Averyt, K., Clemmer, S., and Rogers, J.: The Water Implications of Generating Electricity: Water Use across the United States Based on Different Electricity Pathways through 2050, Environ. Res. Lett., 7, 045803, https://doi.org/10.1088/1748-9326/7/4/045803, 2012.
Mekonnen, M. M. and Hoekstra, A. Y.: The green, blue and grey water footprint of animals and animal products. (Value of water research report 48; No. 48), Delft, the Netherlands: Unesco-IHE Institute for Water Education, http://doc.utwente.nl/76912, 2010.
Mekonnen, M. M., Gerbens-Leenes, P. W., and Hoekstra, A. Y.: The Consumptive Water Footprint of Electricity and Heat: A Global Assessment, Environmental Science: Water Research & Technology, 1, 285–297, https://doi.org/10.1039/c5ew00026b, 2015.
Mielke, E., Anadon, L. D., and Narayanamurti, V.: Water Consumption of Energy Resource Extraction, Processing, and Conversion, Belfer Center for Science and International Affairs, http://live.belfercenter.org/files/ETIP-DP-2010-15-final-2.pdf (last access: 13 October 2017), 2010.
Nationella emissionsdatabasen: RUS, Nationella Emissionsdatabasen, http://www.airviro.smhi.se/cgi-bin/RUS/apub.html_rusreport.cgi, last access: 29 May 2017.
Nilsson, M., Griggs, D., and Visbeck, M.: Policy: Map the Interactions between Sustainable Development Goals, Nature News, 534, 320, https://doi.org/10.1038/534320a, 2016.
Pacetti, T., Lombardi, L., and Federici, G.: Water–energy Nexus: A Case of Biogas Production from Energy Crops Evaluated by Water Footprint and Life Cycle Assessment (LCA) Methods, J. Clean. Prod., 101, 278–291, https://doi.org/10.1016/j.jclepro.2015.03.084, 2015.
Region Gotland: Gotland i Siffror 2016, Gotland i Siffror, Visby, Region Gotland, available at: http://gotland.se/statistik (last access: 29 May 2017), 2016.
Rio Carrillo, A. M. and Frei, C.: Water: A Key Resource in Energy Production, Energ. Policy, 37, 4303–4312, https://doi.org/10.1016/j.enpol.2009.05.074, 2009.
Sachs, J., Schmidt-Traub, G., Kroll, C., Durand-Delacre, D., and Teksoz, K.: SDG Index & Dashboards – Global Report, New York, Bertelsmann Stiftung and Sustainable Development Solutions Network (SDSN), available at: https://issuu.com/unsdsn/docs/sdg_index_dashboard_full (last access: 29 May 2017), 2016.
SCB Statistikdatabasen: SCB Statistikdatabasen [Electronic resource] Statistical database, available at: http://www.scb.se/Statistikdatabasen, last access: 29 May 2017.
Scott, C. A., Pierce, S. A., Pasqualetti, M. J., Jones, A. L., Montz, B. E., and Hoover, J. H.: Policy and Institutional Dimensions of the Water–energy Nexus, Energ. Policy, Sustainability of biofuels, 39, 6622–6630, https://doi.org/10.1016/j.enpol.2011.08.013, 2011.

Scown, C. D., Horvath, A., and McKone, T. E.: Water Footprint of U.S. Transportation Fuels, Environ. Sci. Technol., 45, 2541–2553, https://doi.org/10.1021/es102633h, 2011.

Spang, E. S., Moomaw, W. R., Gallagher, K. S., Kirshen, P. H., and Marks, D. H.: The Water Consumption of Energy Production: An International Comparison, Environ. Res. Lett., 9, 105002, https://doi.org/10.1088/1748-9326/9/10/105002, 2014.

Stokes, J. and Horvath, A.: Life-Cycle Assessment of Urban Water Provision: Tool and Case Study in California, J. Infrastruct. Syst., 17, 15–24, https://doi.org/10.1061/(ASCE)IS.1943-555X.0000036, 2010.

Theobald, E. J., Ettinger, A. K., Burgess, H. K., DeBey, L. B., Schmidt, N. R., Froehlich, H. E., Wagner, C., HilleRisLambers, J., Tewksbury, J., Harsch, M. A., and Parrish, J. K.: Global Change and Local Solutions: Tapping the Unrealized Potential of Citizen Science for Biodiversity Research, Biol. Conserv., 181, 236–244, https://doi.org/10.1016/j.biocon.2014.10.021, 2015.

Tukker, A.: Life Cycle Assessment as a Tool in Environmental Impact Assessment, Environ. Impact Assess., 20, 435–456, https://doi.org/10.1016/S0195-9255(99)00045-1, 2000.

Vattenfall, A. B.: Certified Environmental Product Declaration EPD® of Electricity from Vattenfall Nordic Nuclear Power Plants, Certified Environmental Product Declaration EPD® S-P-00923. UNCPC Code 17, Group 171, available at: http://www.environdec.com/en/Detail/epd923 (last access: 10 October 2017), 2016.

Wamsler, C., Luederitz, C., and Brink, E.: Local Levers for Change: Mainstreaming Ecosystem-Based Adaptation into Municipal Planning to Foster Sustainability Transitions, Global Environ. Chang., 29, 189–201, https://doi.org/10.1016/j.gloenvcha.2014.09.008, 2014.

Xue, B. and Tobias, M.: Sustainability in China: Bridging Global Knowledge with Local Action, Sustainability, 7, 3714–3720, https://doi.org/10.3390/su7043714, 2015.

Yang, Z., Dong, W., Wei, T., Fu, Y., Cui, X., Moore, J., and Chou, J.: Constructing Long-Term (1948–2011) Consumption-Based Emissions Inventories, J. Clean. Prod., 103, 793–800, https://doi.org/10.1016/j.jclepro.2014.03.053, 2015.

Proc. IAHS, 376, 35–43, 2018
https://doi.org/10.5194/piahs-376-35-2018

Open Access
PIAHS

Water, energy and agricultural landuse trends at Shiroro hydropower station and environs

Olubunmi Adegun, Olalekan Ajayi, Gbolahan Badru, and Shakirudeen Odunuga
Department of Geography, University of Lagos, Lagos, Nigeria
Correspondence: Olubunmi Adegun (oadegun@unilag.edu.ng)

Received: 14 June 2017 – Accepted: 26 July 2017 – Published: 1 February 2018

Abstract. The study examines the interplay among water resources, hydropower generation and agricultural landuse at the Shiroro hydropower station and its environs, in north-central Nigeria. Non-parametric trend analysis, hydropower footprint estimation, reservoir performance analysis, change detection analysis, and inferential statistics were combined to study the water-energy and food security nexus. Results of Mann–Kendall test and Sen's slope estimator for the period 1960 to 2013 showed a declining rainfall trend at Jos, around River Kaduna headwaters at $-2.6\,\mathrm{mm\,yr^{-1}}$, while rainfall at Kaduna and Minna upstream and downstream of the reservoir respectively showed no trend. Estimates of hydropower footprint varied between 130.4 and 704.1 $\mathrm{m^3\,GJ^{-1}}$ between 1995 and 2013. Power generation reliability and resilience of the reservoir was 31.6 and 38.5 % respectively with year 2011 being the most vulnerable and least satisfactory. In addition to poor reliability and resilience indices, other challenges militating against good performance of hydropower generation includes population growth and climate change issues as exemplified in the downward trend observed at the headwaters. Water inflow and power generation shows a weak positive relationship with correlation coefficient (r) of 0.48, indicating less than optimal power generation. Total area of land cultivated increased from 884.59 $\mathrm{km^2}$ in 1986 prior to the commissioning of the hydropower station to 1730.83 $\mathrm{km^2}$ in 2016 which signifies an increased contribution of the dam to ensuring food security. The reality of reducing upstream rainfall amount coupled with high water footprint of electricity from the reservoir, therefore requires that a long term roadmap to improve operational coordination and management have to be put in place.

1 Introduction

The interdependence of water, food and energy are widely recognised as important drivers of socioeconomic development. The importance of these natural resources are embedded in the United Nations (UN) 2030 Agenda for Sustainable Development Goals (SDG) 2, 6, 7 (UN SDG Report, 2016; ICSU-ISSC, 2015). With Nigeria's population expected to increase to 262, 599 000 by 2030 (UNDESA, 2015), the country faces a great challenge of meeting these goals. For example, energy demand in the country averages 7 % per year, with an estimated increase in demand for years 2000 and 2012 which is put at 14 and 37 TWh respectively, and estimates for 2020, 2030 and 2040 put at 68, 146 and 291 TWh. Similarly, this challenge may become more glaring with more than 90 million people not having access to grid electricity in Nigeria, yet the widespread use of back-up generators has further reduced the number of people without access to any form of electricity significantly (IEA, 2014).

The link between the food-water-energy nexus and climate, especially rainfall in the study area greatly influences water availability, agricultural production, and electricity generation and supply, in the long run. For instance, high inter-annual variability and marked seasonality of rainfall, coupled with episodes of streamflow drought, have often resulted in reduced power production of hydropower stations in the country owing to low gauge height level, while excessive rainfall sometimes results in downstream flooding with dire consequences especially for the cultivated lands.

Furthermore, the location of the country's hydropower stations in the savannah region suggests that a high volume of water will be lost through evaporation hence the need to know the water footprint of these reservoirs.

Published by Copernicus Publications on behalf of the International Association of Hydrological Sciences.

A number of approaches have been used to estimate the footprint of hydropower reservoirs in different parts of the world. Bueno et al. (2016) used the conventional gross water footprint method for the Camargos Hydropower Reservoir in Brazil, and reported that the average hydropower footprint in Brazil exceeded the global average. Herath et al. (2011) estimated the hydropower footprint of all major hydropower reservoirs in New Zealand using a combination of the consumptive water use method, the net consumptive water use method, and the net water balance method.

Zhao and Liu (2015) developed a method that utilises an allocation coefficient to allocate water footprint among the different ecosystem services offered by the Three Gorges Reservoir in China. The study also compared the estimates of the water footprint obtained with the values derived from estimation based on the gross water footprint method. The result of the study showed that the gross water footprint method over-estimated the hydropower footprint. Mekonen and Hoekstra (2012) employed the gross water footprint method to estimate the hydropower footprint of 35 selected hydropower reservoirs in different parts of the world. The study revealed that hydropower reservoirs located in Ghana, Mozambique, Zambia, Zimbabwe, and Kenya all had high hydropower footprint, with Akosobo-Kpong reservoirs having the highest value globally.

In evaluating the performance of water resources systems, Hashimoto et al. (1982) used three indices; reliability, resilience and vulnerability to evaluate the performance of a water resources system. The indices have also been used to assess water resources systems performance under different prevailing natural and operational conditions. For example, Longobardi et al. (2014) used the 3 indices to assess the performance of a multi-purpose dam in southern Italy under conditions of climate variability. Gohraian et al. (2016) used the 3 performance indices in combination with a hydrological model, a water system model and climate change projections in a study of the Salt Lake City's water supply system under climate change conditions. In a study by Raje and Mujumdar (2010) the reliability index was used to evaluate the performance of a reservoir in Orissa, India, with respect to hydropower generation, irrigation and flood control. The indices of resiliency and vulnerability were used to assess the system's performance in relation to hydropower generation only.

With respect to the performance of the Shiroro Hydropower facility as a water resources system, only a few studies exist in the literature. This include a study by Sule (1988), in which the future operations and performance of the reservoir system which was still under construction were simulated under different operating scenarios using a combination of Probabilistic Dynamic Programming Optimisation Model and the indices of reliability, resilience and vulnerability. The result of the scenario simulation showed severe and frequent failures at the 70 % plant factor recommended for Nigeria. Based on this result, it was concluded that reservoir operations should be carried out at 40 to 50 % power plant factor. Similarly, a recent study by Gbadamosi et al. (2015), which analysed the performance and efficiency of the Shiroro Hydropower Station using the plant availability factor, the capacity factor and the overall efficiency factor, showed that the capacity factor of the hydropower system varied between 23.5 and 50.7 % compared with internationally benchmark of 80 %.

With regards to the estimation of water footprint in the studies highlighted above, hydropower generation was shown as a large consumer of water, and the method, a veritable indicator of resource use. The general output from the review of studies on water resources systems performance indicated that the 3 indices are useful tools in the description of the frequency and duration of system failure and recovery. The studies also showcase how the indices can serve as a basis for the improvement of the operations and management of different water resources infrastructures and systems.

Considering the importance of the Shiroro hydropower station in the water-energy and its food security interconnections within the River Kaduna catchment, and indeed the county at large, there is a need to have a better understanding of the drivers and feedback mechanisms at play. To achieve this therefore, this study utilises a combination of statistical analyses, hydropower water footprint estimation based on the gross water footprint method, water resources systems performance indices evaluation and change detection analysis.

2 Study area

The Shiroro Hydropower reservoir is a storage based hydroelectric facility (Fig. 1) located in the Shiroro Gorge approximately between latitude 9°46′35.29″ and 10°08′36.65″ N and longitude 6°50′51.23″ and 6°53′14.53″ N. It is located approximately 90 km southwest of Kaduna, on River Dinya. The facility has an installed capacity of 600 MW (Kolo, 1996). The reservoir has a surface area of about 320 km^2, a maximum length of 32 m and a total storage capacity of 7 billion m^3 (Suleiman and Ifabiyi, 2015).

About 70 % of inflows into the reservoir is from River Kaduna, with lateral contributions from Rivers Dinya, Sarkin-Pawa, Erena and Muyi (Adie et al., 2012; Eze, 2006). The reservoir is located in the River Kaduna catchment in the Guinea Savanna zone of the country, between the semi-arid climate of the north and the sub-humid climate of the south. The climate of the catchment is consistent with the rest of the country. Rainfall in the area is controlled by the tropical maritime (mT) air mass. During the dry season the area is dominated by the tropical continental (cT) air mass (Anyadike, 1993; Iloeje, 2009). The dry season is between the month of November and March, while the rainy season commences in April or May and lasts till October, (Jimoh and Ayodeji, 2003). Average annual rainfall in the catchment is about 1204.91 mm, with an average of about 110 days per

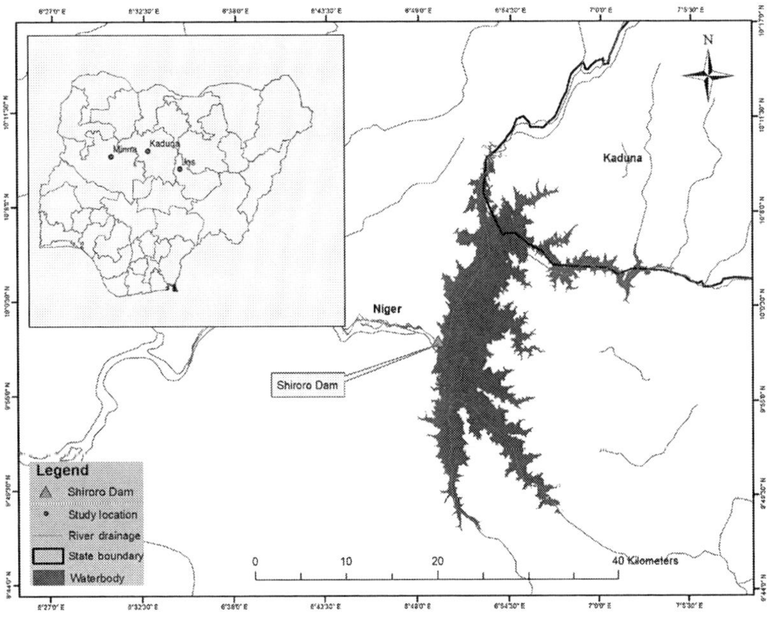

Figure 1. Location of Shiroro hydropower dam.

year receiving a rainfall amount of 0.1 mm (Okafor et al., 2017). Annual temperature around the reservoir varies between 27 and 35 °C (Suleiman and Ifabiyi, 2015).

3 Methodology

Historical monthly rainfall data for locations within the catchment, Jos, Kaduna and Minna for 1960 to 2013 was obtained from the Nigerian Meteorological Agency (NIMET). Monthly data on inflow, turbine discharge, reservoir evaporation and energy generated for 1995 to 2013 was obtained from the Shiroro Hydropower Station.

To establish the presence or absence of trend in rainfall within the catchment, a Mann–Kendall trend test was carried out in XLSTAT. The Mann–Kendall test statistic is expressed as:

$$Z_c \begin{cases} \dfrac{S-1}{\sqrt{\mathrm{var}(S)}}, & S > 0 \\ 0 & S = 0 \\ \dfrac{S+1}{\sqrt{var(S)}}, & S < 0 \end{cases} \tag{1}$$

in which

$$S = \sum_{i=1}^{n-1} \sum_{k=i+1}^{n} \mathrm{sgn}(x_k - x_i) \tag{2}$$

where x_k, x_j are the sequential data values, n is the length data set, and $\mathrm{sgn}(\theta)$ is equal to 1, 0, -1 if θ is greater than, equal to, or less than zero, respectively. The Sen's slope estimator is expressed as:

$$\beta = \mathrm{Median}\left(\frac{x_i - x_j}{i - j}\right), \quad \forall j < i \tag{3}$$

in which $1 < j < I < n$. β is the median overall combination of the record pairs for the whole dataset, and therefore resistant to extreme observation or outliers. A positive value of the β indicates a slope of an upward trend and a negative value represents a downward trend (Machiwal and Jha, 2012).

The water footprint of electricity from hydropower generation was estimated based on Mekonnen and Hoekstra (2012). This is expressed as:

$$\mathrm{WF} = \frac{\mathrm{WE}}{\mathrm{EG}}. \tag{4}$$

The total volume of water evaporated (WE, $m^3\,yr^{-1}$) from the reservoir on an annual basis is expressed in Eq. (5) as:

$$\mathrm{WE} = \left(10 \times \sum_{t-1}^{365} E\right) \times A. \tag{5}$$

The system performance evaluation of the hydropower station was based on Hashimoto et al. (1982). The 3 indices of system performance, reliability, resilience and vulnerability are stated in Eq. (6)–(8) respectively as:

$$r_t = 1 - \frac{f_p}{n} 0 \le r_t \le 1, \quad f_p \le n \tag{6}$$

Table 1. Data sources and characteristics.

Data	Year	Instrument	Resolution	Source
LANDSAT TM	1985	Thematic Mapper 7, 4, 2 bands	30 m	USGS, glovis.org
LANDSAT ETM	2002	Enhanced TM-7, 4, 2 bands	30 m	
LANDSAT ETM	2016	OLI-7, 5, 3 bands	30 m	

Table 2. Result of Mann–Kendall trend test.

Location	Kendall's Tau	P-value	Level of significance	Sen's slope (mm yr^{-1})
Jos	−0.185	0.048	0.05	−2.609
Kaduna	−0.102	0.279	0.05	−0.824
Minna	−0.049	0.601	0.05	−1.554

where f_p is the number of failure periods out of a total of n periods.

The resilience of a system denoted by Υ_{mean} describes how quickly a system is likely to recover from an unsatisfactory state or failure event. If d_j is the duration of each failure or unsatisfactory state, then resilience can be expressed as:

$$\left[\frac{1}{N}\sum_{j=1}^{N} d_j\right] - 1. \quad (7)$$

Vulnerability V_{mean} is a measure of the extent of failure which in this study is the maximum value of shortfall among the values of unsatisfactory period. If X_t is the system performance variable (annual inflow for power generation) and X_0 is the threshold value (mean long term inflow for power generation), the value for each unsatisfactory period can be computed as:

$$V_j \sum_{t=1}^{d_j} [X_t - X_0] \quad (8)$$

and vulnerability (V_{mean}) calculated according to Kjelsen and Rosberg (2004) as:

$$V_{mean} = \frac{1}{N}\sum_{j=i}^{N} v_j. \quad (9)$$

The degree of relationship between inflow into the reservoir and power generated, and between rainfall and reservoir inflow was determined using Pearson correlation coefficient, expressed as:

$$r = \sum xy / N\sigma_x\sigma_y \quad (10)$$

where r is the correlation coefficient, x and y are data variables under consideration, and σ_x and σ_y are standard deviation of the variables under consideration.

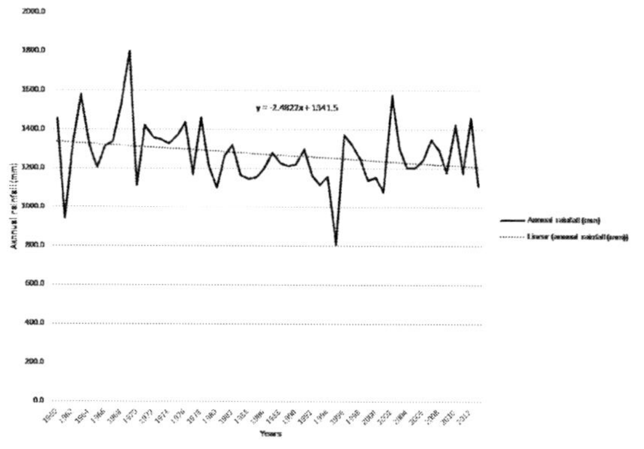

Figure 2. Trend of annual rainfall at Jos.

The USGS Landuse Classification model was used to determine the landuse change characteristics around the Shiroro Dam in three progressive scenarios: 1986, 2000 and 2016. This is done through the remote sensing technique (Table 1) using analytical tool of ENVI. This landuse adopted a pragmatic hierarchical land use classification scheme where seven major landuse classes were identified. About 1000 pixels of training dataset were randomly sampled from the spectral signature of each of the classes to define their respective land use/land cover type. In addition, the Maximum Likelihood classifier was used to extract these classes of landuse. Analysis was based on area calculation for the three static years (1986–2000 and 2000–2016).

4 Results and discussion

4.1 Rainfall trend and rainfall-reservoir inflow relationship

The results of the Mann–Kendall trend test for the rainfall trend (Table 2) shows that the computed p-value for Jos (Fig. 2) is lower than the significance level, $\alpha = 0.05$, therefore the alternative hypothesis (H_1) of the existence of trend in the time series was accepted. The result of the Sen's slope estimator (Q) also indicates a statistically significant downward trend at a declining rate of -2.6 mm yr^{-1}. This shows that rainfall around the headwaters of River Kaduna at Kangimi Hills in the Jos Plateau has been experiencing a declining trend. According to Olaniran (2002), negative rain-

Table 3. Annual water footprint of Shiroro hydropower reservoir between 1995 and 2013.

Year	1995	1996	1997	1998	1999	2000	2001	2002
WFP ($m^3 GJ^{-1}$)	668.0	431.6	NA*	384.6	NA*	468.1	328.4	489.1
Year	2003	2004	2005	2006	2007	2008	2009	2010
WFP ($m^3 GJ^{-1}$)	368.9	340.6	704.1	130.4	320.5	348.9	341.2	353.6
Year	2011	2012	2013					
WFP ($m^3 GJ^{-1}$)	297.0	281.5	269.3					

NA* = not available.

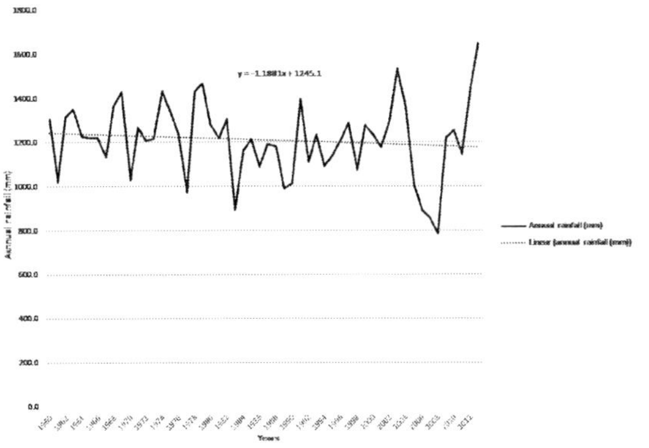

Figure 3. Trend of annual rainfall at Kaduna.

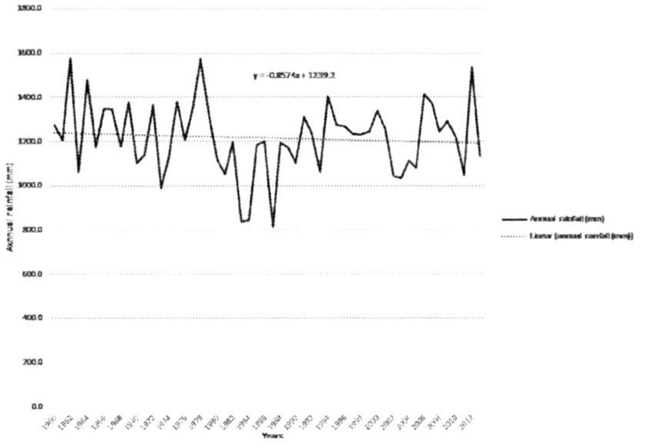

Figure 4. Trend of annual rainfall at Minna.

fall trends in parts of Nigeria is owing to an early retreat of rainfall and a decline in the number of rain days at the end of the rainy season. The implication of this is a gradual reduction of surface runoff and discharge of River Kaduna over the course of the study period. This can consequently lead to the inability of the hydropower reservoir to be filled to its maximum pool elevation because of the reduced inflow contribution to the reservoir, leading to the disruption of normal reservoir operations and hence hydropower generation (Oladipo, 1994). Declining rainfall trend upstream of the dam also means that the volume of tail race discharges available for downstream uses, especially in the dry season will be significantly reduced.

For Kaduna and Minna (Figs. 3 and 4) representing locations along the middle and lower course of the River, the computed p-value was greater than the level of significance, hence the acceptance of the null hypothesis (H_0) of none existence of trend in the time series. The results indicate a weak and statistically insignificant trend over time. The Q values of -1.55 and $-0.8\,mm\,yr^{-1}$ for Kaduna and Minna respectively also suggest that the declining annual rainfall is not statistically significant. The generally declining annual rainfall at the headwater region portends negative consequences for hydropower generation, water supply, rain-fed and irrigated agriculture, and hence food security.

The result of the Karl Pearson correlation coefficient ($r = 0.85$) for catchment rainfall–reservoir inflow relationship shows a high degree of positive relationship. This results shows the strong linkage that exist between rainfall in the catchment and inflow into the reservoir. This suggests that the lower the amount of rainfall available, the lesser the reservoir inflow and stage height, and the lesser the capacity of the capacity of installed turbines to generate power.

4.2 Hydropower water footprint

The water footprint of electricity from the hydropower station for the period 1995 to 2013, (Table 3) varied between $130.4\,m^3\,GJ^{-1}$ (year 2006) and $704.1\,m^3\,GJ^{-1}$ (year 2005), while the average footprint was $383.91\,m^3\,GJ^{-1}$. This implies that the highest consumptive use of water via evaporation from the Shiroro reservoir occurred in 2005, while the

Table 4. Landuse change analysis.

	1986		2000		2016	
Landuse	Area (km^2)	%	Area (km^2)	%	Area (km^2)	%
Built-up	211.93	4.2	232.95	4.8	252.01	5.5
Sandy shore	1.07	0.5	0.7	0.3	0.25	0.1
Fallow land	1709.9	37.6	1541.46	33.9	1069.75	23.6
Farming	884.59	19.5	1142.03	25.2	1730.83	38.1
Ridge	439.93	9.7	438.63	9.7	438.14	9.6
Savanna/Orchard bush	1044.65	23	934.09	20.6	798.01	17.6
Waterbody	248.73	5.5	250.74	5.5	251.61	5.5
Total	4540.6	100	4540.6	100	4540.6	100

Source: Author's GIS Analysis.

lowest consumptive use occurred the following year, 2006. The coefficient of variation of the water footprint and evaporation was 36 and 25 % respectively, while the coefficient of variation for electricity generated was 15 %. This suggests a moderate inter-annual variability in the water footprint, evaporation and energy generated from the hydropower station. Water footprint could not be computed for 1997 and 1999 due to the non-availability of data on electricity generated in both years.

As a large consumer of water, it is necessary for the decision makers to have an adequate knowledge of the water footprint of hydropower generation. This will help to forestall undesirable consequences the on downstream environmental flows and ensure sustainable management for electricity generation and other uses.

4.3 Hydropower reservoir performance and inflow–power generation relationship

The result of the reliability and resilience of the hydropower reservoir with respect to power generation are 31.6 and 38.5 % respectively. This is an indication that the hydropower power system was less than reliable more than half of the time, which is also reflected in the low level of resilience as suggested by the low resilience value of 38.5 %. The system was at its most vulnerable and least satisfactory in 2011, with an annual inflow of 112 000 m^3. The result of the Pearson correlation coefficient ($r = 0.48$) shows a weak degree of positive relationship between inflow into the reservoir and power generated.

From these results it is obvious that the hydropower system performance is below the optimal level. However, the low performance of the hydropower system does not signify a lack of water in the reservoir, rather other operational deficiencies also contributed to the below optimal level of performance. These operational deficiencies include breakdown of equipment, inadequate maintenance, low level of inflow into the reservoirs, poor coordination and inefficient operating guidelines (Sule, 1988). In view of the current state of affairs, there is a need for an efficient reservoir operational strategy to reduce the system's vulnerability and improve the level of reliability and resilience.

Table 5. Statistics of area and yield of crops cultivated in Niger State (2011–2016).

	Maize	Rice
Minimum area (10^3 HA)	150.5	264.0
Maximum area (10^3 HA)	355.3	513.6
Average area (10^3 HA)	215.9	506.3
Minimum yield (10^3 MT HA^{-1})	2.2	2.0
Maximum yield (10^3 MT HA^{-1})	3.0	5.3
Average yield (10^3 MT HA^{-1})	2.6	3.7

Source: Niger State Agricultural and Mechanisation Development Authority (2017).

4.4 Landuse changes analysis

Table 4 and Figs. 5 to 7 show the static distribution of the landuse/landcover pattern in three periodical scenarios of landuse dynamics around the Shiroro Dam Development Project in 1986, 2000 and 2016 respectively. Subsistence land area for crop production increased from 884.59 km^2 in 1986 to 1730.83 km^2 in 2016. The significant gains in agricultural landuse in the area can be attributed to the construction of the dam, which although has no designed irrigation scheme but serves as a source of irrigation water for peasant farmers around the hydropower reservoir through self- effort. Crops cultivated in the area include yam, maize, guinea, corn, sugar cane, millet, beniseed, groundnut, rice, banana, plantain and cotton. Although there are no official records of the hectares of land devoted to the cultivation of these crops in the area, records for the whole of Niger State (Table 5), where the hydropower reservoir is located, shows that the area devoted to the cultivation of maize between 2011 and 2016 varied between 150.5 and 355.3 thousand hectares, with an average area of 215.9 thousand hectares while the area devoted to

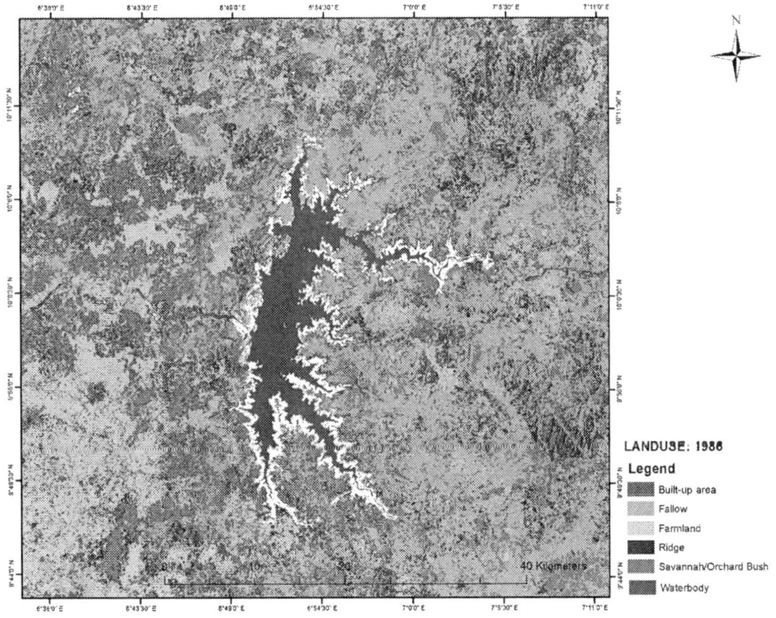

Figure 5. Landuse/landcover pattern in 1986.

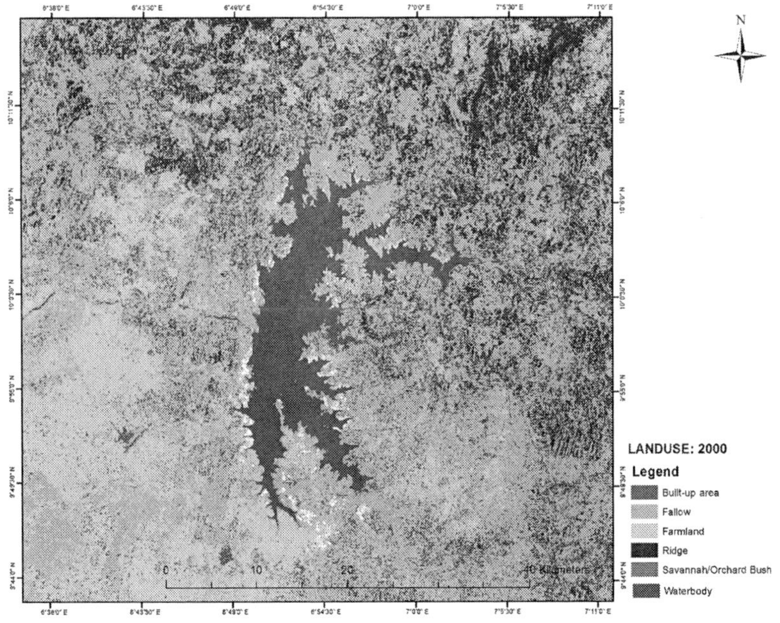

Figure 6. Landuse/landcover pattern in 2000.

rice varied between 264.0 and 513.6 thousand hectares, with an average areal extent of 506 thousand hectares. Maize yield for the same period varied between 2.2 and 3.0 thousand metric tonnes per hectare, with an average of 2.6 thousand metric tonnes per hectare, while rice yield varied between 2.0

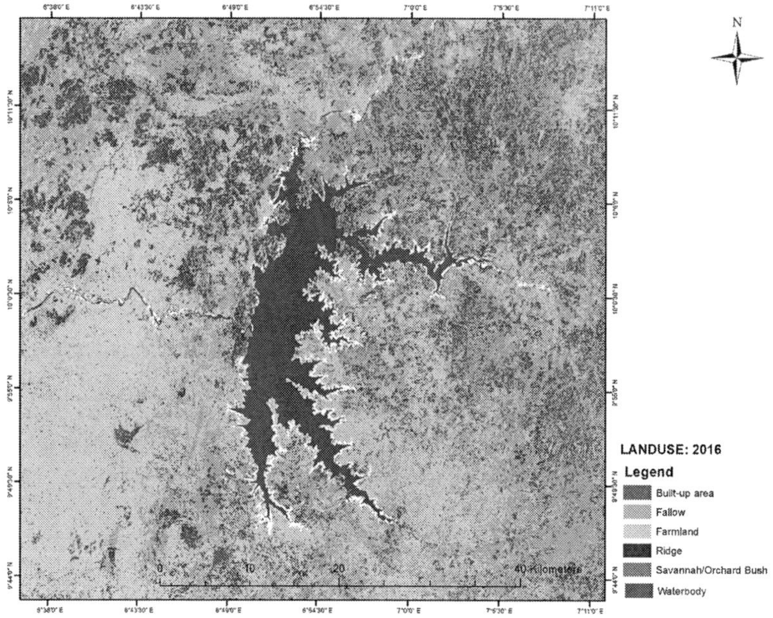

Figure 7. Landuse/landcover pattern in 1986.

and 5.3 thousand metric tonnes per hectare, with an average yield of 3.7 thousand metric tonnes per hectare.

Looking at the emergence of the hydropower facility and the trend of agricultural landuse in the area, the inter-play of water, energy and agricultural nexus in Shiroro can be deemed to have had a positive impact on the livelihood of the local populace and food security in the area and the country at large.

5 Summary and conclusion

The study showed how a combination of methods can be used to investigate feedback mechanisms operational at the food-water-energy nexus at the Shiroro hydropower station and its environs. Although the results of the reliability and resilience of the hydropower reservoir with respect to hydropower generation are below the average performance with less than 40 % for all indices. The comparative gain in agricultural sector through water utilisation by peasant farmers actually provide an additional gain derived from the inter-play of food-water-energy nexus. In the face of a reducing upstream rainfall amount coupled the with high water footprint of electricity from the reservoir, it is recommended that a long term roadmap to improve the operational coordination and management has to be put in place. Also, a formal irrigation scheme should be provided to strengthen agricultural production in the area.

Data availability. Links to the USGS Satellite Imagery data is available at Earth explorer http://earthexplorer.usgs.gov. The hydro-meteorological and agricultural data is not publicly accessible.

Author contributions. OAd conceptualized and lead the team on this research. She also wrote the manuscript and effected the corrections during the write-up. OAj gathered and processed all the spatial data used for this article. GB assisted in the gathering of the hydro-meteorological data used for the study. SO reviewed the write-up from time to time for necessarycorrections.

Competing interests. The authors declare that they have no conflict of interest.

Special issue statement. This article is part of the special issue "Water security and the food-water-energy nexus: drivers, responses and feedbacks at local to global scales". It is a result of the IAHS Scientific Assembly 2017, Port Elizabeth, South Africa, 10–14 July 2017.

Acknowledgements. The authors wish to express their profound gratitude to the following institutions. The Nigerian Meteorological Agency for the climatic data, the Shiroro Hydropower Station for the hydrological and power generation data, and the Niger State Agricultural and Mechanisation Development Authority for the agricultural yield data.

Edited by: Barry Croke
Reviewed by: Mykhaylo Romashchenko and one anonymous referee

References

Adie, D. B., Ismail, A., Mohammed, M. M., and Aliyu, U. B.: Analysis of the water resources potential and useful life of Shiroro Dam, Nigeria, Niger. J. Basic Appl. Sci., 20, 341–348, 2012.

Anyadike, R. N. C.: Seasonal and annual rainfall variations over Nigeria, Int. J. Climatol., 13, 567–580, https://doi.org/10.1002/joc.3370130507, 1993.

Bueno, E. D., de Mello, C. R., and Alves, G. J.: Evaporation from Camargos hydropower plant reservoir, Brazil. J. Water Resour., 21, 570–575, 2016.

Eze, J. N.: Vulnerability and adaptation to climatic variability and extremes. A case study of flooding in Niger State, MSc dissertation, School of Geography and Environmental Studies, University of Witwatersrand, Witwatersrand, South Africa, 153 pp., 2006.

Gbadamosi, S. L., Ojo, O., Adedayo, O., and Nnaa, L.: Evaluation of operational efficiency of Shiroro hydroelectric plant in Nigeria, Int. J. Sci. Eng. Investigat., 4, 2251–8843, 2015.

Gohraian, E., Burian, S. J., Bardsley, T., and Strong, C.: Incorporating potential severity into vulnerability assessment of water supply systems under climate change conditions, J. Water Resour. Pl. Manage., 14, 1–12, 2016.

Hashimoto, T., Stedinger, J. R., and Loucks, D. P.: Reliability, resiliency and vulnerability criteria for water resource system performance evaluation, Water Resour. Res., 18, 14–20, 1982.

Herath, I., Deurer, M., Horne, D., Singh, R., and Clothier, B.: The water footprint of hydroelectricity: a methodological comparison from a case study in New Zealand, J. Clean Prod., 19, 1582–1589, 2011.

ICSU-ISSC: Review of the sustainable development goals: the science perspective, Paris International Council for Science, Paris, 88 pp., 2015.

IEA – International Energy Agency: Africa energy outlook. A focus on energy prospects in Sub-Saharan Africa, World Energy Outlook Special Report, Paris, 237 pp., 2014.

Iloeje, N. P.: A New Geography of Nigeria, New Revised Edn., Longman Press, Lagos, Nigeria, 2009.

Jimoh, O. D. and Ayodeji, O. S.: Impact of the Gurara River (Nigeria) interbasin water transfer scheme on the Kaduna River at Shiroro dam, in: Water resources systems: Hydrological risk management and development, edited by: Blosch, G., Franks, S., Kumagai, M., Musiake, K., and Rosbjerg, D., Proceedings of Sapporo Symposium, July 2003, Sapporo, 277–286, 2003.

Kjelsen, T. R. and Rosberg, D.: Choice of reliability, resilience and vulnerability estimates for risk assessments of water resources systems, Hydrolog. Sci. J., 49, 755–767, 2004.

Kolo, R. J.: The assessment of physico-chemical parameters of shiroro lake and its major tributaries, edited by: Eyo, A. A., Proceedings of the 13th annual conference of fisheries society of Nigeria, 3–8 November 1996, New Bussa, Nigeria, 260–268, 1996.

Longobardi, A., Mautone, M., and De Luca, C.: Reservoirs performances under climate variability: a case study, Proceedings of the ICWRS2014, June 2014, Bologna, Italy, 526–531, 2014.

Machiwal, D. and Jha, M. K.: Hydrologic Time Series Analysis: Theory and practice, Capital Publishing, Dordrecht, the Netherlands, 2012.

Mekonnen, M. M. and Hoekstra, A. Y.: The blue water footprint of electricty from hydropower, Hydrol. Earth Syst. Sci., 16, 179–187, https://doi.org/10.5194/hess-16-179-2012, 2012.

Niger State Agricultural and Mechanisation Development: Authority Report, Minna, Nigeria, 2017.

Okafor, G. C., Jimoh, O. D., and Larbi, K. I.: Detecting changes in hydro-climatic variables during the last four decades (1975–2014) on downstream Kaduna river catchment in Nigeria, Atmos. Clim. Sci., 7, 161–175, 2017.

Oladipo, E. O.: An indication of abrupt change of rainfall and its impacts on energy development in Nigeria, Proceedings of the International Workshop on Impacts of Climate Change in Energy Development, Jos, 231–238, 1995.

Olaniran, O. J.: Rainfall anomalies in Nigeria: The contemporary understanding, Inaugural Lecture, University of Ilorin, Ilorin, 2002.

Raje, D. and Mujumdar, P. P.: Reservoir performance under uncertainty in hydrologic impacts of climate change, Adv. Water Resour., 33, 312–326, 2010.

Sule, B. F.: Reservoir operation policies for optimising energy generation at shiroro dam, Water Resour. Manage., 2, 209–219, 1988.

Suleiman, Y. M. and Ifabiyi, I. P.: The role of rainfall variability in reservoir storage management at Shiroro hydropower dam, Nigeria, Monoma Ethiop. J. Sci., 7, 55–63, 2015.

UNDESA: World population prospects: the 2015 revision, key findings and advance tables, New York, 59 pp., 2015.

UN SDG – United Nations, Department of Economic and Social Affairs, Population Divison: The sustainable development goals report, United Nations Report, New York, 52 pp., 2016.

Zhao, D. and Liu, J.: A new approach to assessing the water footprint of hydroelectric power based on allocation of water foot print among reservoir ecosystem services, Phys. Chem. Earth, 79–82, 40–45, 2015.

Proc. IAHS, 376, 45–50, 2018
https://doi.org/10.5194/piahs-376-45-2018

Open Access
PIAHS

An integrated approach to improving rural livelihoods: examples from India and Bangladesh

Barry Croke[1,2], Wendy Merritt[1], Peter Cornish[3], Geoffrey J. Syme[4], and Christian H. Roth[5]

[1]Fenner School of Environment and Society, Australian National University, Canberra, 0200, Australia
[2]Mathematical Sciences Institute, Australian National University, Canberra, 0200, Australia
[3]School of Science and Health, University of Western Sydney, Sydney, 2751, Australia
[4]School of Business and Law, Edith Cowan University, Perth, 6027, Australia
[5]CSIRO Land and Water, Brisbane, 4001, Australia

Correspondence: Barry Croke (barry.croke@anu.edu.au)

Received: 16 June 2017 – Accepted: 2 August 2017 – Published: 1 February 2018

Abstract. This paper presents an overview of work in West Bengal, Andhra Pradesh and SW Bangladesh through a series of projects from 2005 to the present, considering the impact of farming systems, water shed development and/or agricultural intensification on livelihoods in selected rural areas of India and Bangladesh. The projects spanned a range of scales spanning from the village scale ($\sim 1\,\mathrm{km}^2$) to the meso-scale ($\sim 100\,\mathrm{km}^2$), and considered social as well as biophysical aspects. They focused mainly on the food and water part of the food-water-energy nexus. These projects were in collaboration with a range of organisations in India and Bangladesh, including NGOs, universities, and government research organisations and departments. The projects were part funded by the Australian Centre for International Agricultural Research, and built on other projects that have been undertaken within the region. An element of each of these projects was to understand how the hydrological cycle could be managed sustainably to improve agricultural systems and livelihoods of marginal groups. As such, they evaluated appropriate technology that is generally not dependent on high-energy inputs (mechanisation). This includes assessing the availability of water, and identifying potential water resources that have not been developed; understanding current agricultural systems and investigating ways of improving water use efficiency; and understanding social dynamics of the affected communities including the potential opportunities and negative impacts of watershed development and agricultural development.

1 Introduction

Water security for agriculture is a vital component of food security and improving livelihoods. Water security can deliver higher and more stable crop yields, for example through resilience to dry spells in the growing season or it may create new growing seasons, thus intensifying cropping. Improving water security can be achieved across a range of scales. At the village scale (few km^2), focus can be on improving access to water through rainwater harvesting methods that can be applied by farmers. This can include structures that enable access to shallow groundwater systems (e.g. hand dug pits), small ponds or dams that capture local runoff and the use of deep-rooted perennial plants. A key component of such work is education, including learning about alternative crop options, weed control and fertilizer use, as well as training in managing finances. At meso-scale ($\sim 100\,\mathrm{km}^2$), options extend to include larger structures that require collaboration across a village or even between villages. These include the use of surface water retention structures (e.g. larger check dams, wells) that provide extended access to surface water and shallow groundwater as well as increasing groundwater recharge. At a larger scale (e.g. basin or state), institutional and governmental drivers become increasingly important.

Underpinning this work (particularly for the higher rainfall in the village scale study area), was the principle that shallow or "renewable" groundwater is the component of groundwater that can be utilized on a sutainable basis (Pavelic et al., 2015).

The projects discussed here were funded by the Australian Centre for International Agricultural Research (ACIAR). ACIAR invests in research in agricultural landscapes that is intended to deliver positive social, economic and environmental outcomes to a range of beneficiaries, including farmers and rural poor (http://aciar.gov.au/aboutus). Increasingly the agency is funding research projects that move beyond more traditional discipline based research to consider the challenges facing the rural poor and particularly women that prevent them engaging in agricultural activities in a way that meaningfully improves their lives, and to work with communities to identify opportunities into the future.

2 Village scale

An Australian-Indian government co-funded project (Cornish et al., 2013, 2015a, b) explored the watershed development (WSD) techniques being used by an NGO (PRADAN) to improve livelihoods in tribal villages in the undulating East India Plateau (EIP). These techniques included structures that individual farmers could implement including dug wells, pits, ponds or small dams to capture rainfall in the wet season and/or tap into shallow, seasonally recharged groundwater resources, and the use of bunds in poor (local) upland areas to retain runoff for high value, perennial crops (e.g. silviculture for silk production; mango orchards). Such work can be done by individual smallholders with or without mechanisation, or collaboratively between farmers across a village. The emphasis was on providing small, mostly privately-owned distributed water resources within a community-developed and owned WSD plan. To supplement this work, PRADAN also trains villagers, particularly women, to manage their finances better. This includes creating women's self-help groups, and encouraging the villagers to pool resources and to lend money amongst themselves. The aim is to improve their natural resources and management practices without incurring unmanageable debt, often accessing funds from a range of government programs.

The village-scale project focused on two villages in the Purulia district of West Bengal, in the foot hills of the EIP. PRADAN had already engaged in one village (Amagara) for a few years and was planning to start work in the second village (Pogro) after the first year of the project, during which socioeconomic and biophysical baselines were established.

3 Amagara village

PRADAN had undertaken some WSD in Amagara, but there had been no change in agricultural practices or more effective use of water. Therefore, the focus in this village was on participatory development of new crop options and improved management practices to use water more effectively, and evaluating the agricultural, economic and social effectiveness of the work (Cornish et al., 2015b).

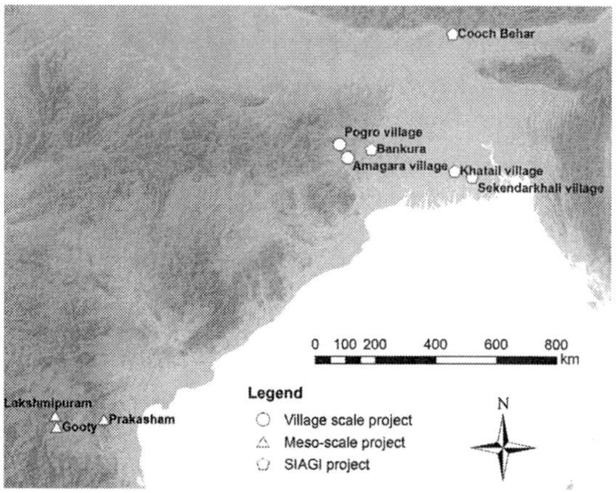

Figure 1. Map showing locations of the study sites for the three projects.

By the end of the project there was clear evidence of village-wide change. Farmers had re-evaluated the worth of "poor" uplands and learnt to use them intensively for vegetables, moved from mono-crop rice to multiple cropping, experimented with different crops to improve their income, and, as a side effect, increase diet diversity (Cornish et al., 2015b). As a result, seasonal migration was significantly reduced, and there was increased re-investment in improving natural resources, and more was spent on education and medical care. The women reported working harder in the fields, but being more engaged in agricultural decision-making and feeling rewarded in terms of both income and respect. Issues around land leasing agreements and water access arose as the value of uplands for vegetables became clear, along with the value of the water resources created by WSD. These issues impacted most on landless families. One adverse effect of the localized work of PRADAN was that during the project, mango orchards were partially burnt by members from a neighbouring village. While a setback, this lead to discussions between the villages, and reinforced the need to avoid creating potential conflict points that may arise through focused WSD work.

Intensive WSD is expensive, but widely thought to improve agricultural production and livelihoods. However, an important finding from the long-term water balance modelling in this project (Cornish et al., 2015a), combined with participatory experiments (Cornish et al., 2015b), was that comprehensive WSD is not essential in East India (which has relatively high rainfall) to make substantial and more immediate improvements in livelihoods. Much can be done to improve the use of rainfall through a variety of agronomic interventions that reduce rice yield variability and increase cropping intensity and productivity without a commensurate increase in risk. Measures can be taken to use more effectively existing water resources (pits, ponds and wells) and similar additional low-cost investments more effectively, to develop horticulture on appropriate uplands, convert risky rice-growing land to other crops, and intensify production in the post-monsoon period.

3.1 Pogro village

The Pogro village is located within a 2 km^2 catchment area (headwater of the Chapai Nula) located near Jaldha and within the Damodar sub-basin of the Lower Ganga Basin. Work here focused on hydrology and agronomy, with a strong social component as in Amagara, involving villagers in experiments and receiving training in the use of fertilizers, weed control and the selection of crops (Cornish et al., 2015b).

Activities in both villages included training villagers to recognize and choose appropriate fertilizers and conducting simple field tests to determine suitable application rates, enabling villagers to observe responses to fertilizers (particularly phosphorous and nitrogen). Villagers also experimented with a range of crop alternatives to rice in trials on all land classes. There was also a gender equity component to the research, focusing on the effectiveness of the self-help groups for women which were created by PRADAN. Ultimately, the effectiveness of the watershed development work hinged on the effectiveness of self-help groups, as these enabled the villagers to develop the capacity to continue development after the end of the project (as observed in Amagara).

3.1.1 Hydrology

The Pogro village area was the focus of the hydrological component of this project. The average annual rainfall is approximately 1300 mm, dominated by the southwest monsoon. The outlet of the catchment is defined by a railway culvert, with a small adjoining catchment (approximately 0.5 km^2) used as a reference. As the culverts were also a thoroughfare for vehicles, people and animals, a v-shaped speed hump design was used (in addition to the culvert structure) as the flow control. An automatic logger was used to monitor water levels in each culvert. An automatic weather station was used to monitor rainfall, temperature, relative humidity wind speed and solar radiation at a 1 min resolution. This was supplemented by manual recording of rainfall, temperature and relative humidity, as well as water levels in ponds, wells and piezometers installed across the catchment (Croke et al., 2008).

The data were used to understand the hydrological response of the study site and evaluate the water balance across the study period (2006–2011), though logged water level data in the culverts were only available from 2009 following permission to install equipment in the railway culvert. This information was used to develop guidelines for the design of watershed development plans, as well as a soil water balance model and micro-catchment scale hydrological model (Cornish et al., 2013, 2015a).

4 Meso-scale

Defining the meso-scale to be about 100–200 km^2 (or about the size of a Mandal or Block – the second lowest level of Indian government), this project (Syme et al., 2015; Reddy and Syme, 2015) focused more on the social side of the issue of water shed development, while also having a strong focus on the groundwater system. Indian project partners included the Livelihoods and Natural Resources Management Institute, National Geophysical Research Institute, Central Research Institute for Dryland Agriculture, and the International Water Management Institute.

This project had a greater focus on social aspects, with a number of surveys conducted to gain understanding of the attitudes in the study areas, as well as the available resources and limitations. The information collected in these surveys was used to develop a Bayesian network model for each study area (Merritt et al., 2011, 2016).

The study sites were the Vajralavanka-Maruvavanka catchment (approximately 150 km^2) in the Anantapur and Kurnool districts, and the Peethuruvagu catchment (approximately 100 km^2) in the Prakasham district in Andhra Pradesh. In comparison to the Pogro study site, the study sites in Andhra Pradesh have a semi-arid climate with an average annual rainfall of approximately 500–700 mm, approximately half that of the study site in West Bengal, with rain from the southwest monsoon being less reliable. As a result, there is a greater reliance on groundwater resources in the area relative to West Bengal.

The micro-catchment scale hydrological model developed in the previous project was used as the starting point for developing a meso-catchment scale model. The primary study sites were not gauged, so the model was developed using data for a nearby gauged site (Lakshmipuram) with a catchment area of approximately 2750 km^2 (an order of magnitude larger than the study sites), and the model used to predict the available surface and shallow groundwater water resource in the study sites.

5 Recommendations from village and meso-scale projects

Based on the findings in these two projects, a number of recommendations were put forward for future WSD planning and implementation work, as well as policies related to WSD. A selection of these is given here (for more detail see Cornish et al. (2013, 2015a, b), Syme et al. (2015) and Reddy and Syme (2015).

5.1 Village-scale project

- In East India, proven agronomic interventions that reduce risks with rice and increase crop productivity and cropping intensity should be scaled out, especially where there is not the funding or technical capacity to implement WSD.
- Training farmers in basic agriculture and to effectively conduct their own field experiments is vital to ensuring WSD is effective in positively influencing livelihoods, because providing additional water resources alone does not necessarily lead to effective use of those (or existing) water resources. Extension agencies should adopt a flexible approach capable of handling weather/climate changes; and take a participatory approach allowing farmers (including wife) to learn rather than be told.
- Farmers will develop their own adaptive systems, but they need to be equipped to understand their water resources and match these to crop options.
- In the relatively high rainfall area in East India, research should strive towards climate-responsive farming systems with water requirements that broadly match water supply from current rainfall, including water captured in the annually recharged shallow groundwater system as well as local surface runoff, thus minimising dependency on deep groundwater for irrigation.

5.2 Meso-scale project

- The present blanket, fixed per unit area allocation of funds to WSD is not appropriate. Rather, allocation needs to be location-specific with funds divided, within and between watersheds depending on agro-climatic and hydrogeological factors.
- It is not necessary for the entire area to be treated. In fact, some portion of the watershed could be left untreated as a buffer (donor) for the rest of the watershed.
- Household resilience is an important indicator of the socio-economic impacts of WSD at all scales, and needs to be explicitly considered.
- Achieving equity remains a difficult issue to be resolved at all scales, and demands judicious planning of interventions with justice principles in mind.
- As competition for land and water increases, at all scales there will need to be improved regulation of agreements that give marginal families access to land and water.
- Bayesian Networks (BNs) have the potential to support integrated watershed management program (IWMP) design and evaluation.
- Stakeholder engagement is an integral part of any research project aiming for policy changes.
- The project has helped to improve the Detailed Project Report (DPR) process - the vehicle by which the IWMP is implemented while there has been some progress in institutional terms understanding process evaluation is still in its early stages.
- Similarly, continued efforts are needed to build capacity within the GOs and NGOs in the implementation of the DPR.
- In terms of the hydrological and biophysical research it is necessary that the impact of these factors is examined in the context of different landscapes, particularly in alluvial river basins such as the Ganges.

6 Multi-scale climate change adaptation project

An important consideration for water and food security is the impact of projected climate change, and developing capacity to adapt to such changes. In addition to the aforementioned projects, ACIAR funded a project (the ACCA project) entitled "Developing multi-scale climate change adaptation strategies for farming communities in Cambodia, Lao PDR, Bangladesh and India" (Van Wensveen et al., 2016; Hochman et al., 2017a, b). Using a participatory on-farm research process coupled to modelling, this project developed and tested a range of options to respond to climate risk, centred on strategic irrigation. Farmer capacity to adapt to possible climate change was built through Climate Clubs and village level Climate Information Centres.

7 Current research: increased emphasis on social research and community engagement

Drawing on learnings from the all of the above projects, the "Promoting socially inclusive and sustainable agricultural intensification in West Bengal and Bangladesh" project (SIAGI) is a collaborative project involving research and NGO partners from India, Bangladesh, Australia and Wageningen University (https://siagi.org/about-siagi/). It aims to further understand how socially disadvantaged groups, such as landless, marginal smallholders or women-headed household engage with agriculture and to work with these households (and the broader community) to identify and promote opportunities for these households to benefit from agricultural intensification. The project is undertaking integrated research into nutrition sensitive agriculture, value chains and pro-poor markets and social justice and equity issues.

In Bangladesh, SIAGI is working in two villages in two polders in the coastal zone of Bangladesh. In this region, activities such as cropping, livestock rearing and aquaculture are constrained or enhanced by many climatic or environmental factors that vary in time and space: some key drivers being soil, groundwater and surface water salinity in the rabi season (in particular), with associated freshwater shortages for agriculture and consumption, and drainage and waterlogging issues during and at the end of the monsoon. Salinity and a lack of available water currently contribute to much of the land being fallow in the dry season. The biophysical conditions in the polders are influenced to a considerable extent by the maintenance and management of polders and the irrigation infrastructure (canals and sluice gates). Social processes around who uses the water, for what, and when (and the equity of decisions made within a community) are a key focus of SIAGI. SIAGI is partnering with the "Cropping system intensification in the salt-affected coastal zones of Bangladesh and West Bengal, India" project, which is undertaking soil, water and crop research in the same study villages as well as investigating polder-scale water and salt management strategies (http://aciar.gov.au/project/lwr/2014/073).

8 Conclusions

Water and food security are vital for improving livelihoods in disadvantaged rural areas. Hydrological understanding is central to the design of watershed development works intended to improve water security. However, whether new water resources are made available through WSD or not, it is important to address the design and management of farming systems to ensure water is used productively and sustainably. It is also vital for social constraints and equity to be considered. Finally, a participatory learning approach is useful for ensuring that any watershed development work is effective in producing long-term improvement in livelihoods.

Data availability. The dataset for the village scale project is publicly available.

Competing interests. The authors declare that they have no conflict of interest.

Special issue statement. This article is part of the special issue "Water security and the food-water-energy nexus: drivers, responses and feedbacks at local to global scales". It is a result of the IAHS Scientific Assembly 2017, Port Elizabeth, South Africa, 10–14 July 2017.

Acknowledgements. The authors acknowledge funding from the Australian Centre for International Agricultural Research for projects: Water harvesting and better cropping systems for smallholders of the East India Plateau (LWR/2002/100), Impacts of meso-scale Watershed Development (WSD) in Andhra Pradesh (India) and their implications for designing and implementing improved WSD policies and programs (LWR/2006/072), and Promoting socially inclusive and sustainable agricultural intensification in West Bengal and Bangladesh (LWR/2014/072).

Edited by: Graham Jewitt
Reviewed by: Nadja den Besten and one anonymous referee

References

Cornish, P., Croke, B., Kumar, S., and Karmakar, D.: Water harvesting and better cropping systems for smallholders of the East India Plateau, Final report for LWR/2002/100, available at: http://aciar.gov.au/publication/FR2013-25 (last access: 15 September 2017), 2013.

Cornish, P. S., Choudhury, A., Kumar, A., Das, S., Kumbakhar, K., Norrish, S., and Kumar, S.: Improving crop production for food security and improved livelihoods on the East India Plateau II, Crop options, alternative cropping systems and capacity building, Agr. Syst., 137, 180–190, https://doi.org/10.1016/j.agsy.2015.02.011, 2015a.

Cornish, P. S., Karmakar, D., Kumar, A., Das, S., and Croke, B.: Improving crop production for food security and improved livelihoods on the East India Plateau. I, Rainfall-related risks with rice and opportunities for improved cropping systems, Agricultural Systems, 137, 166–179, https://doi.org/10.1016/j.agsy.2015.01.008, 2015b.

Croke, B., Norrish, N., Kumar, A., Islam, A., Dey, P., Cornish, P., Kumar, S., Ghosh, J., Insights into catchment behaviour for water harvesting assessment in the East India Plateau, in: Proceedings of the iEMSs Fourth Biennial Meeting: International Congress on Environmental Modelling and Software (iEMSs 2008), edited by: Sànchez-Marrè, M., Béjar, J., Comas, J., Rizzoli, A. E., and Guariso, G., International Environmental Modelling and Software Society, Barcelona, Catalonia, July 2008, 1, 454–461, 2008.

Croke, B., Islam, A., and Cornish, P.: Hydrological data for Pogro study site, Purulia district, West Bengal, India, HydroShare, https://doi.org/10.4211/hs.8e0ad0e45df44404b0ad42ecc84a95da, 2017.

Hochman, Z., Horan, H., Reddy, D. R., Sreenivas, G., Tallapragada, C., Adusumilli, R., Gaydon, D., Kokic, P., Singh, K. K., and Roth, C. H.: Smallholder farmers managing climate risk in India: 1. adapting to a variable climate, Agr. Syst., 150, 54–66, 2017a.

Hochman, Z., Horan, H., Reddy, D. R., Sreenivas, G., Tallapragada, C., Adusumilli, R., Gaydon, D., Laing, A., Kokic, P., Singh, K. K., and Roth, C. H.: Smallholder farmers managing climate risk in India: 2. is it climate-smart?, Agr. Syst., 151, 61–72, 2017b.

Merritt, W. S., Reddy, V. R., Rao, K. V., Pavelic, P., Ahmed, S., Ranjan, R., Croke, B. F. W., and Syme, G. J.: Integrated modelling for understanding watershed development impacts on social and biophysical systems, in: 19th International Congress on Modelling and Simulation (MODSIM), edited by: Chan, F., Marinova, D., and Anderssen, R. S., 2887–2893, 12–16 December 2011, Perth, Australia, available at: http://mssanz.org.au/modsim2011/G3/merritt.pdf, 2011.

Merritt, W. S., Patch, B., Reddy, V. R., and Syme, G.: Modelling livelihoods and household resilience to droughts using Bayesian networks, Environ. Dev. Sustain., 18, 315–346, https://doi.org/10.1007/s10668-015-9650-1, 2016.

Pavelic, P., Senaratna Sellamuttu, S., Johnston, R., McCartney, M., Sotoukee, T., Balasubramanya, S., Suhardiman, D., Lacombe, G., Douangsavanh, S., Joffre, O., Latt, K., Zan, A. K., Thein, K., Myint, A., Cho, C., and Htut, Y. T.: Integrated assessment of groundwater use for improving livelihoods in the dry zone of Myanmar, Colombo, Sri Lanka: International Water Management Institute (IWMI), IWMI Research Report 164, 47 pp., https://doi.org/10.5337/2015.216, 2015.

Syme, G. J., Reddy, V. R., Pavelic, P., Croke, B., and Ranjan, R.: Confronting scale in watershed development in India, Hydrogeol. J., 20, 985–993, https://doi.org/10.1007/s10040-011-0824-0, 2012.

Syme, G., Reddy, V. R., Ahmed, S., Rao, K. V., Pavelic, P., Croke, B., Merritt, W., Ranjan, R., Tallapragada, C., Sreedevi, P. D., Sarah, S., Nallan, S., Rout, S., Guillaume, J., and Bernet, D.: Impacts of meso-scale Watershed Development (WSD) in Andhra Pradesh (India) and their implications for designing and implementing improved WSD policies and programs, Final report for LWR/2006/072, available at: http://aciar.gov.au/publication/fr2015-05 (last access: 15 September 2017), 2015.

Reddy, V. R. and Syme, G. J.: Integrated Assessment of Scale Impacts of Watershed Intervention: Assessing Hydrogeological and Bio-physical Influences on Livelihoods, Elsevier, Amsterdam, 420 pp., 2015.

Van Wensveen, M., Williams, L. M., and Roth, C. H.: Developing multi-scale adaptation strategies for farming communities in Cambodia, Laos, Bangladesh, India. ACIAR Final Report for LWR/2008/019, available at: http://aciar.gov.au/publication/fr2016-10 (last access: 15 September 2017), 2016.

Proc. IAHS, 376, 51–55, 2018
https://doi.org/10.5194/piahs-376-51-2018

Open Access
PIAHS

Forecasting domestic water demand in the Haihe river basin under changing environment

Xiao-Jun Wang[1,2,3], Jian-Yun Zhang[1,2], Shamsuddin Shahid[4], Yu-Xuan Xie[5], and Xu Zhang[1,2]

[1]State Key Laboratory of Hydrology-Water Resources and Hydraulic Engineering, Nanjing Hydraulic Research Institute, 210029 Nanjing, China
[2]Research Center for Climate Change, Ministry of Water Resources, 210029 Nanjing, China
[3]State Key Laboratory of Water Environment Simulation, School of Environment, Beijing Normal University, 100875 Beijing, China
[4]Faculty of Civil Engineering, Universiti Teknologi Malaysia (UTM), 81310 Johor Bahru, Malaysia
[5]Haihe River Water Conservancy Commission, 300170 Tianjin, China

Correspondence: Xiao-Jun Wang (xjwang@nhri.cn)

Received: 8 June 2017 – Accepted: 25 July 2017 – Published: 1 February 2018

Abstract. A statistical model has been developed for forecasting domestic water demand in Haihe river basin of China due to population growth, technological advances and climate change. Historical records of domestic water use, climate, population and urbanization are used for the development of model. An ensemble of seven general circulation models (GCMs) namely, BCC-CSM1-1, BNU-ESM, CNRM-CM5, GISS-E2-R, MIROC-ESM, PI-ESM-LR, MRI-CGCM3 were used for the projection of climate and the changes in water demand in the Haihe River basin under Representative Concentration Pathways (RCPs) 4.5. The results showed that domestic water demand in different sub-basins of the Haihe river basin will gradually increase due to continuous increase of population and rise in temperature. It is projected to increase maximum 136.22×10^8 m^3 by GCM BNU-ESM and the minimum 107.25×10^8 m^3 by CNRM-CM5 in 2030. In spite of uncertainty in projection, it can be remarked that climate change and population growth would cause increase in water demand and consequently, reduce the gap between water supply and demand, which eventually aggravate the condition of existing water stress in the basin. Water demand management should be emphasized for adaptation to ever increasing water demand and mitigation of the impacts of environmental changes.

1 Introduction

Residential water demand is defined as the amount of water withdrawal for residential purposes such as in-house water use for drinking, food preparing, bathing, clothes and dishes washing, toilet flushing, etc. as well as outdoor water needs for gardening, lawn watering, etc. (Blokker et al., 2010; Wang et al., 2017a). Water demand is different from water consumption which indicates the portion of total water use that is not returned to the original water source. Generally, residential water withdrawal or demand are measured in household or regional levels and forecasted. The domestic water demand depends on number of factors including population growth, socio-economic development, climate change, urbanization, water saving technological advances, water tariff, etc. (Sebri, 2014; Wang et al., 2016). A number of recent studies assessed the impacts of those factors on domestic water demand in different regions (Dursun, 2010; Zachariadis, 2010; Protopapas et al., 2010; Karamouz et al., 2011; Jakimavičius and Kriaučiūnienė, 2013; Browne et al., 2013; Wang et al., 2014, 2017b; Babel et al., 2014; Price et al., 2014; Walker et al., 2016; Froelich and Magiera, 2016). The studies revealed that the World will face a growing challenge to maintain safe and adequate water supplies due to continuous growth of population and changes in climate. Residential water demand shares a major portion of total water demand in urbanized catchment. Therefore, understanding possible changes in residential water demand due to environmental changes is important for water resources planning and man-

Published by Copernicus Publications on behalf of the International Association of Hydrological Sciences.

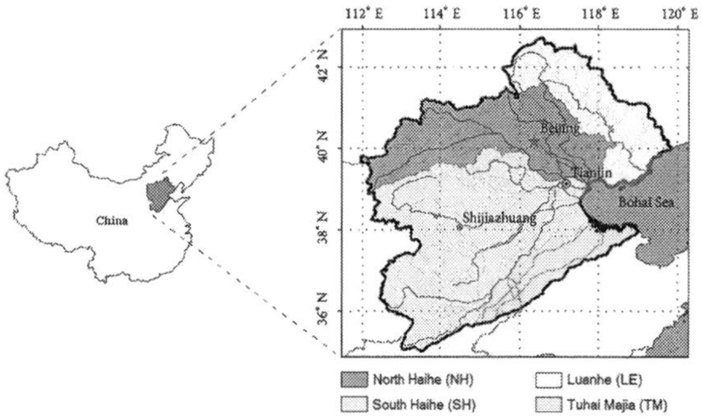

Figure 1. Location of Haihe River basin in the map of China. The sub-basins of the river and the location of major cities in the basin are also shown.

agement and adaptation to environment changes (Shahid et al., 2016).

The objective of this study is to develop a statistical model for forecasting domestic water demand due to climate change, population growth and technological advances. The proposed model is developed through the analysis of the effects of climate variables on domestic water use, and then applied in Haihe River basin of China for forecasting water demand. The Haihe River (112–120° E and 35–43° N), flows through the major cities like Beijing and Tianjin (Fig. 1) plays a major role in national, economic, and social development of China. The Haihe River basin covers 8 % of total area of China (318 000 km^2), but about 10 % of total population reside in the basin and about 13 % of total GDP comes from the basin. About 45 % area of the basin is urbanized. High population density, rapid urbanization and socio-economic development have caused a sharp increase in water demand, particularly residential water demand. This has made the basin one of the most water stressed region of China. However, water security in the basin has extreme importance in China's economic development and societal safety (Liu and Speed, 2009; Wang et al., 2015). Therefore, forecasting residential water demand is very important for the basin.

2 Materials and methods

2.1 Data and sources

The population, domestic water use and urbanization rate data for the period 1980–2012 were collected from Chinese Water Resources Bulletins (MWR, 2000–2012). The recent data of domestic water use in four sub-basins of Haihe River basin, namely, the Luanhe River and Coastal areas in Eastern Hebei (LE), Northern Haihe (NH), Sorthern Haihe Region (SH), Tuhai Majiahe Region (TM) as well as for the entire Haihe River basin (HR) were collected from the Haihe River Water Resources Bulletins (HRCC, 2001–2012).

The bias corrected daily temperature data of the basin for the period of 1961–2012 were obtained from National Climate Center of China (NCCC). The bias corrected future projections of daily average temperature by an ensemble of 7 GCMs namely BCC-CSM1-1, BNU-ESM, BNU-ESM, CNRM-CM5, GISS-E2-R, MIROC-ESM, PI-ESM-LR, and MRI-CGCM3 under the Representative Concentration Pathways (RCPs) 4.5 were also collected from NCCC.

2.2 Modeling domestic water demand

According to the water resources planning technical specifications in China, domestic water demand is forecasted by multiplying the projected population (10^4 P) with the projected per capita water consumption (L^{-1} p^{-1} day^{-1}; Zhang, 2005; Wang et al., 2017b). The advances in water saving technologies, water pricing, possible changes in people's lifestyles, etc. are considered in projection of per capita water consumption. However, it does not consider the influence of climate change in water demand. We attempted to improve the exiting method of water demand forecasting by incorporating climate change factor.

The most common approach for considering the effect of climatic conditions on water demand is to include temperature and precipitation variables in the regression equations as there is a common conception that higher temperature and less rainfall make people consume more residential water. However, the influence of rainfall on domestic water consumption is often found negligible (Slavíková et al., 2013; Zhou, 2016). Zhou (2016) found the residential water demand elasticity of rainfall is insignificant for China, while it is significantly positive (0.4299) in case of daily average temperature. Slavíkoná et al. (2013) also reported no impact of rainfall on residential water demand in Czech Republic. Therefore, only the impact of temperature rise due to climate change is considered to modify the existing equation of residential water demand forecasting. The changes in residential water demand $\Delta \mathrm{LW}^t$ can be expressed as:

$$\Delta \mathrm{LW}^t = d_T \cdot \Delta T \cdot \mathrm{LW}^t \quad (1)$$

Where, LW^t is the residential water demand (10^4 m^3) forecasted using existing method; ΔT is the change in temperature (°C) due to climate change; d_T is the temperature elasticity of domestic water demand, which is defined as,

$$d_T = \frac{\Delta \mathrm{LW}}{\Delta T} \quad (2)$$

Where, ΔLW is the changes in residential water demand in 10^4 m^3 due to changes in temperature (ΔT) in °C.

In this paper, a linear regression model was developed with domestic water use data as independent and temperature as dependent variable in order to estimate the temperature elasticity of domestic water demand. The regression model was developed separately for each sub-basin for the estimation of d_T of each sub-basin.

Table 1. Daily average temperature projected by different GCMs in the Haihe River basin.

	BCC-CSM1-1	BNU-ESM	CNRM-CM5	GISS-E2-R	MIROC-ESM	MPI-ESM-LR	MRI-CGCM3
Base year (1961–2000)	9.36	9.38	9.65	9.54	9.44	9.17	9.33
2020	11.26	10.63	10.16	9.85	11.08	10.04	10.64
2030	10.53	11.59	10.27	10.86	10.57	10.64	9.98

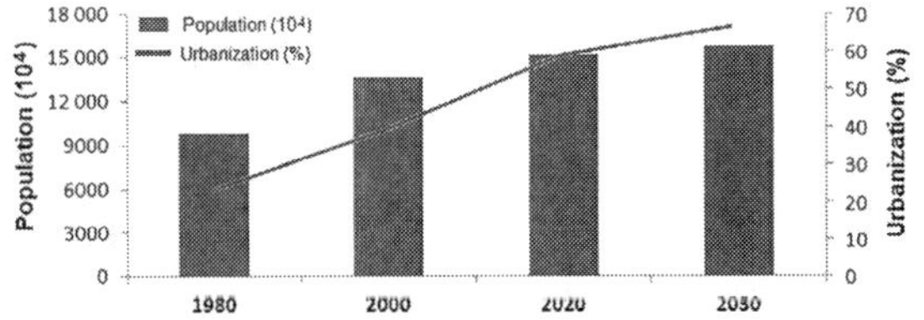

Figure 2. The historical changes and future projections of population and urbanization in the Haihe River basin.

3 Projection of domestic water demand in Haihe River basin

3.1 Projections of population and urbanization

With rapid urbanization, the population in Haihe River basin has grown rapidly in last four decades. The historical changes and future projections of population and urbanization in Haihe River basin is shown in Fig. 2. The basin had a population of 9721×10^4 and only 23.6 % of the basin was urbanized in 1980. The population reached to $13\,606 \times 10^4$ and 42.4 % of the basin became urbanized in 2006. The growing population caused a rapid increase in domestic water demand. For example, the domestic water demand in the basin increased from $1.92 \times 10^8\,m^3$ in 1980 to $5.93 \times 10^8\,m^3$ in 2012. According to the future plan of the river basin, the total population will reach to $15\,117 \times 10^4$ and the urbanized area will reach to 58.6 % in 2020. In 2030, the total population will reach to $15\,752 \times 10^4$ and the urbanized area will reach to 66.4 %.

3.2 Temperature projections

The simulated daily average temperature for the base period (1961–1990) and for the years 2020 and 2030 by 7 GCMs in Haihe River basin are shown in Table 1. The table shows increase in temperature by 0.31 to 1.9 °C in year 2020 and by 0.65 to 2.19 °C in year 2030 in the basin. The highest increase was projected by BCC-CSM1-1 in 2020 and BNU-ESM1 in 2030, while the lowest by GISS-E2-R in 2020 and CNRM-CM5 in 2030.

3.3 Projection of domestic water demand

The water resources planning technical specifications in China estimated per capita water consumption considering the advances in water saving technologies, water tariff, changes in socio-economic condition, etc. Urbanization (shown in Fig. 2) is considered as proxy for socio-economic condition. The per capita water consumptions projected for year 2020 and 2030 were multiplied with projected population (shown in Fig. 2) to estimate the amount of residential water demand due to population growth, water saving technological advances and socio-economic development. The temperature elasticity of water demand and temperature rises projected by different GCMs were then used to estimate the domestic water demand using Eq. (1) in order to incorporate climate change factor in water demand. The water demand projected by different GCMs for different sub-catchments as well as the whole Haihe River basin for years 2020 and 2030 are presented in Fig. 3.

The figure shows that domestic water demand in Haihe River basin will continuously increase due to population growth and temperature rise. The results were found to vary for different GCMs. The BCC-CSM1-1 projected the highest increase in residential water demand in the basin ($110.37 \times 10^8\,m^3$) in 2020, while the GISS-E2-R projected the lowest increase ($85.82 \times 10^8\,m^3$). In 2030, the BNU-ESM projected the highest increase in domestic water demand ($136.22 \times 10^8\,m^3$), while the CNRM-CM5 projected the lowest increase ($107.25 \times 10^8\,m^3$). The results indicate large uncertainty in projection of domestic water demand in the Haihe River basin under projected scenarios.

The changes in water demand were also found to vary for different sub-basins. The highest increase was projected in the SH sub-basin, followed by NH. The SH and NH are the most populated regions of the Haihe River basin. Therefore, it can be anticipated that increased water demand will accelerate the condition of growing water stress in the basin.

4 Discussion and conclusion

A model has been developed for forecasting domestic water demand in the context of climate change, population growth and technological development. The model is applied in the Haihe River basin of China which is considered as one of the most economically and politically impor-

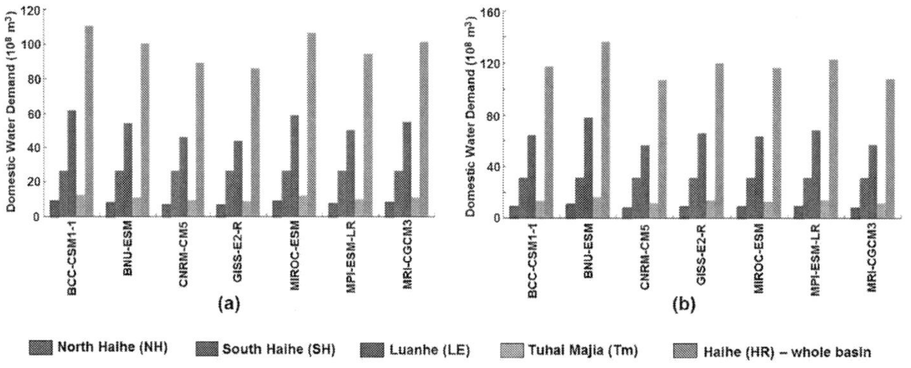

Figure 3. Projection of domestic water demand in different sub-basins of the Haihe River in years **(a)** 2020; and **(b)** 2030.

tant but worst water stress region of the country. The results show considerable variations in domestic water demand with space, time and scenarios. Overall, all the GCMs reveal increase in water demand with time, mainly due to increase in population and rise in temperature. The domestic water demand in the Haihe River basin is projected in the range of 85.82×10^8 to 110.37×10^8 m^3 in 2020, and 107.25×10^8 to 136.22×10^8 m^3 in 2030. Considerable variations in projections of domestic water demand by different GCMs indicate large uncertainty in projections. A large variation in water demand among different sub-basins indicates heterogeneous change in water demand. The higher increase in water demand might be in the South and North sub-basins, which are already under water stress due to high population density and rapid urbanization. Large increase in domestic water demand in those regions would aggravate the condition of existing water stress. The study indicates prudent management of water resources in the Haihe River basin is very important for mitigation of the impacts of environmental changes. It is expected that the knowledge generated in this paper on possible future changes in domestic water demand would help in water resources planning and management of the basin. The method proposed in this article will help in rapid assessment of water demand in the context of changing scenarios. However, it should be noted that the method presented in this paper is only one of a new innovation in the field of forecasting domestic water demand under changing environment, which need further attention to improve. The domestic water demand is influenced by many factors which can be considered in future research for better projection of water demand. It is also required to quantify the uncertainty in water demand due to population growth and socio-economic development.

Data availability. Data are available in the Chinese Water Resources Bulletins (2000–2012) and the Haihe River Water Resources Bulletins (1998–2012).

Competing interests. The authors declare that they have no conflict of interest.

Special issue statement. This article is part of the special issue "Water security and the food–water–energy nexus: drivers, responses and feedbacks at local to global scales". It is a result of the IAHS Scientific Assembly 2017, Port Elizabeth, South Africa, 10–14 July 2017.

Acknowledgements. We are grateful to the National Key R&D Program of China (NO: 2017YFC0403506), National Natural Science Foundation of China (No. 51309155), Strategic Consulting Projects of Chinese Academy of Engineering (NO: 2016-ZD-08-05-02), China Water Resource Conservation and Protection Project (No. 126302001000150001) and Open Research Fund of State Key Laboratory of Simulation and Regulation of Water Cycle in River Basin China Institute of Water Resources and Hydropower Research) (NO: IWHR-SKL-201515) for providing financial support for this research. We are also thankful to reviewers (Mykhaylo Romashchenko and another anonymous referee) and the editor Barry Croke for their helpful comments and suggestions.

Edited by: Barry Croke
Reviewed by: Mykhaylo Romashchenko
and one anonymous referee

References

Babel, M. S., Maporn, N., and Shinde, V. R.: Incorporating Future Climatic and Socioeconomic Variables in Water Demand Forecasting: A Case Study in Bangkok, Water Resour. Manage., 28, 2049–2062, 2014.

Blokker, E. J. M., Vreeburg, J. H. G., and van Dijk, J. C.: Simulating residential water demand with a stochastic end-use model, J. Water Res. Pl.-ASCE, 136, 19–26, 2010.

Browne, A. L., Medd, W., and Anderson, B: Developing Novel Approaches to Tracking Domestic Water Demand Under Uncertainty – A Reflection on the "Up Scaling" of Social Science Approaches in the United Kingdom, Water Resour. Manage., 27, 1013, https://doi.org/10.1007/s11269-012-0117-y, 2013.

Dursun, S.: Effect of Global Climate Change on Water Balance of Beyşehir Lake (Konya – Turkey), BALWOIS 2010, Ohrid, Republic of Macedonia, 25–29 May 2010.

Froelich, W. and Magiera, E.: Forecasting Domestic Water Consumption Using Bayesian Model, Intelligent Decision Technologies 2016, 337–346, 2016.

HRCC: Haihe River water resources bulletin, Haihe River Conservancy Commission, Tianjin, 2001–2012.

Jakimavičius, D. and Kriaučiūnienė, J.: The climate change impact on the water balance of the Curonian Lagoon, Water Resour., 40, 120–132, 2013.

Karamouz, M., Yazdi, M. S. S., Ahmadi, B., and Zahraie, B.: A system dynamics approach to economic assessment of water supply and demand strategies. Paper presented at the World Environmental and Water Resources Congress 2011: Bearing Knowledge for Sustainability, Proceedings of the 2011 World Environmental and Water Resources Congress, 1194–1203, 2011.

Liu, B. and Speed, R.: Water resources management in the People's Republic of China, Int. J. Water Resour. Dev., 25, 193–208, 2009.

MWR: China Water Resources Bulletin, Ministry of Water Resources (MWR), Beijing, 2000–2012.

Price, J. I., Chermak, J. M., and Felardo, J.: Low-flow appliances and household water demand: an evaluation of demand-side management policy in Albuquerque, New Mexico, J. Environ. Manage., 133, 37–44, 2014.

Protopapas, A. L., Katchamart, S., and Platonova, A.: Weather effects on daily water use in New York City, J. Hydrol. Eng., 5, 332–338, 2000.

Sebri, M.: A meta-analysis of residential water demand studies, Environ. Dev. Sustain., 16, 499–520, 2014.

Shahid, S., Wang, X.-J., Harun, S. B., Shamsudin, S. B., Ismail, T., and Minhans, A.: Climate variability and changes in the major cities of Bangladesh: observations, possible impacts and adaptation, Reg. Environ. Change, 16, 459–471, 2016.

Slavíková, L., Malý, V., Rost, M., Petružela, L., and Vojáček, O.: Impacts of Climate Variables on Residential Water Consumption in the Czech Republic, Water Resour. Manage., 27, 365–379, 2013.

Walker, D., Creaco, E., Vamvakeridou-Lyroudia, L., Farmani, R., Kapelan, Z., and Savic, D.: Forecasting Domestic Water Consumption from Smart Meter Readings Using Statistical Methods and Artificial Neural Networks, Procedia Engineering, 119, 1419–1428, 2016.

Wang, X.-J., Zhang, J.-Y., Wang, J.-H., He, R.-M., ElMahdi, A., Liu, J.-H., Wang, X.-G., King, D., and Shahid, S.: Climate change and water resources management in Tuwei river basin of Northwest China, Mitigation and adaptation strategies for global change, 19, 107–120, 2014.

Wang, X.-J., Zhang, J.-Y., Yang, Z., Shahid, S., He, R.-M., Xia, X.-H., and Liu, J.-H.: Historic water consumptions and future management strategies for Haihe River basin of Northern China, Mitigation and adaptation strategies for global change, 20, 371–387, 2015.

Wang, X.-J., Zhang, J.-Y., Shahid, S., Guan, E., Wu, Y., Gao, J., and He, R.-M.: Adaptation to climate change impacts on water demand. Mitigation and Adaptation Strategies for Global Change, 21, 81–99, 2016.

Wang, X.-J., Zhang, J.-Y., Shahid, S., Xie, W., Du, C.-Y., Shang, X.-C., and Zhang, X.: Modeling domestic water demand in Huaihe River Basin of China under climate change and population dynamics, Environment, Development and Sustainability, online first, 1–14, https://doi.org/10.1007/s10668-017-9919-7, 2017a.

Wang, X.-J., Zhang, J.-Y., Shahid, S., Oyang, R.-L., Guan, T.-S., Xue, J.-G., and Zhang, X.: Impacts of climate variability and changes on domestic water use in the Yellow River Basin of China, Mitigation and Adaptation Strategies for Global Change, 22, 595–608, 2017b.

Zachariadis, T.: Residential water scarcity in Cyprus: impact of climate change and policy options, Water, 2, 788–814, 2010.

Zhang, H.: Strategic study for water management in China, Southeast University Press, Nanjing, 2005.

Zhou, M.: Residential Water Demand in China: Applications of Double-Log Model and EDM System, The Southern Agricultural Economics Association's 2016 Annual Meeting, San Antonio, Texas, 6–9 February 2016.

Proc. IAHS, 376, 57–62, 2018
https://doi.org/10.5194/piahs-376-57-2018

Open Access
PIAHS

A half-baked solution: drivers of water crises in Mexico

Jonatan Godinez Madrigal[1,2], **Pieter van der Zaag**[1,2], **and Nora van Cauwenbergh**[1]

[1]Department of Integrated Water Systems and Governance, IHE-Delft, Delft, 2601 DA, the Netherlands
[2]Department of Water Management, TU Delft, Delft, 2628 CN, the Netherlands

Correspondence: Jonatan Godinez Madrigal (j.godinezmadrigal@un-ihe.org)

Received: 7 June 2017 – Accepted: 4 August 2017 – Published: 1 February 2018

Abstract. Mexico is considered a regional economic and political powerhouse because of the size of its economy, and a large population in constant growth. However, this same growth accompanied by management and governance failures are causing several water crises across the country. The paper aims at identifying and analyzing the drivers of water crises. Water authorities seem to focus solely on large infrastructural schemes to counter the looming water crises, but fail to structure a set of policies for the improvement of management and governance institutions. The paper concludes with the implications of a business-as-usual policy based on infrastructure for solving water problems, which include a non-compliance to the human right to water and sanitation, ecosystem collapses and water conflicts.

1 Introduction

Mexico is a rich and diverse country with a multitude of ecosystems, ranging from water-rich jungles to dry deserts. The country's economy is among the top ten in the world, and second in Latin America. Socio-economic dynamics and water distribution are interlinked, and vary across the country, with a mainly agricultural south, with as much as 11 768 $m^3\,cap^{-1}\,yr^{-1}$ of water availability, producing 20 % of the GDP and with a 23 % of the population, and a heavily industrialized center and north with a low water availability of 1650 $m^3\,cap^{-1}\,yr^{-1}$ and 80 % of GDP and 77 % of the population (Conagua-Semarnat, 2016). Regardless of the south having high water availability, still close to 20 % of the population lack water supply and sanitation, while the semi-arid north almost achieves 100 % coverage (Gobierno de Mexico, 2014). After a 2014 reform to the constitution granting the human right to water and sanitation, the State must guarantee water supply and sanitation to all the population through the local government. To comply with this human right is a daunting task without putting more pressure to water resources, as the current population surpasses 119 million people (INEGI, 2015), which by 2050 is expected to grow to 150 million (Conapo, 2017b). Another pressure on water resources comes with another commitment from the State, which is to reduce poverty, as currently 55 million Mexicans are being considered poor (Coneval, 2017). Recently, the Mexican Government has focused on structural reforms to improve economic growth – along with its concomitant water use, as the only way to put people out of poverty (Gobierno de Mexico, 2017).

These socio-economic tendencies along with the geographical distribution of water has severely strained the water resources in general, and unevenly affected the population. Mexico is approaching several water crises that are recognized by some government institutions, by civil society and scholars (COMDA et al., 2017; Conagua-Semarnat, 2014; Peña-Ramirez, 2012). The most serious four are (1) the growing gap between a sustainable supply and the unrelenting demand, (2) the growing pollution of aquatic ecosystems, (3) urban/rural water access crises and (4) vulnerability under climate change, as described in more detail in Table 1. In order to face these challenges, Conagua (National Water Commission) has embarked on the development of large infrastructural projects like dams, interbasin water transfers and water treatment plants under public-private partnerships. However, following Madani (2014) on the underlying causes of water crisis, water agencies tend to focus only on a "symptom-based management paradigm", rather than addressing the drivers of water crises, defined as the root of the water problems characterized by its dynamism and pervasiveness embedded in the social system. Conagua is the central authority that administers water in Mexico including infrastructure development, quality and quantity water monitoring, water distribution, and issuing of water rights. There are also river basin councils across the country, but Conagua's weight in decision-making still is overwhelming.

The main purpose of this paper is to analyze the water crises in Mexico and its concomitant drivers. The paper is based on an extensive literature review on the dynamics of water problems Mexico has faced over the past decade. First, official data and documents from Conagua and other government agencies related to water were analyzed to picture a narrative paired with data. Second, an analysis was undertaken on critical scientific articles and books that analyze in-depth the socio-political context to the water problems first described. This exercise was subsequently enriched and extended with field work during the first half of 2017 when interviews with water agencies officials and stakeholders were held. The outcome of the analysis of these three sets of data and narratives is the synthesis presented in Table 1.

The structure of the article is as follows: the corpus of the paper analyses each of the four water crises mentioned and their drivers, and then the socio-environmental implications that these crises may carry are discussed, then the paper ends with the concluding remarks.

1.1 Growing gap between water supply and demand

The current gap in the national water balance mounts to 11 500 $hm^3\,yr^{-1}$, as the sustainable water supply (water use that does not compromise ecological flows nor comes from overexploited aquifers) is estimated to be 66 900 $hm^3\,yr^{-1}$ and the water use of all water users is estimated to be 78 400 $hm^3\,yr^{-1}$ (36 % of it being underground water); this gap is only destined to widened in the future, with an expected doubling to 23 000 $hm^3\,yr^{-1}$ (Conagua-Semarnat, 2014). All uses are expected to grow: agriculture, public and industrial. Currently, Mexico has 6 rivers among the world's most depleted freshwater sources, causing high competition between users and depletion of ecological flow (Richter, 2014), the number 7 in the world for most water extracted (Conagua-Semarnat, 2016), and is among the top countries of groundwater consumption that exceeds its recharge (World Economic Forum, 2011), Mexico being the largest groundwater user in Latin America (Scott and Banister, 2008), with 23 % of its aquifers presenting over-exploitation or saline intrusion (Conagua-Semarnat, 2016). The general trend is that water availability per capita will decrease in time as population increases. Water availability in Mexico will drop from more than 5000 $m^3\,cap^{-1}$ in 1990, to less than 3,000 $m^3\,cap^{-1}$ by 2050, whilst Mexico City currently only has a per capita annual water availability of 150 m^3.

Economic growth is promoted regardless of water availability in the region. Decision-making on water allocation rights is still dominated by big players in river basin councils as underground water rights are granted to large food and mining corporations in water scarce regions (COMDA, 2017; Stoltenborg and Boelens, 2016), who use their economic influence to get additional water rights despite drilling bans in place by Conagua, as found during fieldwork in early 2017. Reis (2014) documented how the water authority allowed the existence of the black market in the Valley of Mexico basin, where powerful stakeholders could buy out thousands of cubic meters from impoverished farmers. Although the drilling bans is set in order to recharge over-exploited aquifers, the black market inhibits the sustainable use of them.

Public participation in river basin councils still has no influence on decision-making due to co-opted participation of its sectors' representatives, which represent political interests rather than their constituents' (Mollard et al, 2010). This tilts the debate towards a focus on pre-existing economic interests, and jeopardizes the sustainability of water management. Some attempts have been made to integrate a self-regulation institution aimed at metering underground water extraction and enforcing water use according to water rights to tackle aquifer over-exploitation in closed basins, but it has failed due to political struggles and low user participation (Wester et al., 2009).

Agriculture, the largest water consumer in Mexico, has a great potential to free-up water that can be re-allocated, as at least 14 % of the water allocated to irrigation are water losses that can be avoided. By far the largest part of these avoidable losses occur in the semi-arid north, where a considerable amount of that water evaporates, and thus are considered consumptive losses; avoiding these would really imply net water savings. Despite this situation, the budget to modernize agriculture is smaller than the construction of a single water supply dam (Gobierno de Mexico, 2014).

1.2 Pollution of aquatic ecosystems

Although each year new water treatment plants are built in the country, still almost 70 % of all industrial and 50 % of municipal wastewater and its raw sewage goes untreated into surface bodies of water and the sea (Conagua, 2015). With such numbers it can be assumed that water quality has deteriorated over time; however water quality monitoring has been compromised due to a change in the norms and standards by Conagua in 2002. Therefore there are substantial differences in how water quality performance is recorded, which makes it difficult to assess whether water quality of most water bodies in Mexico has improved or deteriorated over time (Aboites et al., 2008; Hansen and Juárez-Corzo, 2011). This notwithstanding, it is estimated that 20 % of aquifers and over 30 % of surface water are polluted (Conagua-Semarnat, 2016). Currently, there are only three official parameters for monitoring pollution: chemical oxygen demand, biological oxygen demand and total suspended solids. Despite signing the Stockholm convention on Toxic, Persistent and Bioaccumulable Substances (TPBS), Conagua has not monitored TPBS in river basins (Hansen and Juárez-Corzo, 2011).

Table 1. Drivers of water crises in Mexico.

	Growing gap between sustainable water supply and demand	Pollution of aquatic ecosystems	Urban/rural water crisis	Vulnerability under stochastic meteorological events – droughts and floods
Drivers	Population and economic growth	Outdated pollutants regulations	Underfunded water utilities, prone to pork-barrel politics	Flawed metering, monitoring and legal enforcement of water use
	Inefficient agriculture and operation of built infrastructure	Lack of infrastructure and deficient maintenance to leakages and bad water quality	Old distribution system, prone	Chaotic urbanization
	Flawed monitoring metering, and law enforcement	Lack of monitoring and law enforcement – self-regulated industries	Unsustainable and/or conflictive water sources	Over-reliance in large infrastructure for water supply

According to Mexican Law, local governments oversee operation and management of water treatment plants, but operation costs are high, and financial and technical capacity to cover those costs is mostly absent. This situation causes water treatment plants to deteriorate and many fall rapidly into obsolescence (Ramón-Lahera, 2010). This explains why out of a total of $210\,m^3\,s^{-1}$ of waste water produced in the country, only $99.8\,m^3\,s^{-1}$ are treated, despite installed infrastructure with the potential for $140.1\,m^3\,s^{-1}$ if adequately maintained (Gobierno de Mexico, 2014). Even though this model of water treatment is obviously flawed, the main funding for new water treatment plants is allocated exclusively for capital intensive water treatment plants, instead of more passive treatment plants like constructed wetlands or hybrid systems that do not require an overwhelming technical and financial capacity for rural areas or small cities.

The urban model prevalent in Mexican cities makes water treatment complex, as industrial and residential zones are often mixed, thwarting the function of the water treatment plants that are designed either for municipal or industrial wastewater (Ramón-Lahera, 2010). Therefore, most wastewater considered as having been treated does not comply with the standards of receiving water bodies.

Monitoring point source pollution has become a daunting task as resources allocated are far from sufficient. For example, only as few as 214 inspectors are in charge of monitoring close to 475 000 water users that discharge polluted water. With this level of inspection it would take 61 years to check all water users (McCulligh, 2017). To solve this problem, there is a self-regulation policy, without any guarantee of industries reporting real data, and in the case of proven misconduct, industries can easily evade a fine by issuing a complaint on the procedure, which Conagua often accepts, dropping the charges (Ibíd.). Water quality is increasingly becoming an issue dominated by conflict, as most Mexican rivers are polluted, especially by unregulated industries. This situation has evolved into an issue of public health, as many riparian communities near heavily polluted rivers present a surge of skin diseases, kidney failure, and many kinds of cancer (Lezama, 2016).

1.3 Urban/rural water crisis

According to official data, Mexico has achieved a water supply coverage of 92 % of the population. However, this data should be questioned, as it mainly counts infrastructure, not real access to a continuous flow of good quality water, as suggested by COMDA et al. (2017). Water scarcity in major cities is becoming a usual phenomenon (Peña-Ramirez, 2012). Some peri-urban areas in major cities get their water supply through water tank pipes, despite having water connection to a centralized water storage system, as the faucet is dry due to water rationing (Gómez-Valdez and Palerm-Viqueira, 2015). Such situation affects mainly the poorest at the bottom quintile (COMDA et al., 2017). Poor water access can be so critical that it can become a flagship promise for some politicians to get elected. Conditional water supply in exchange for votes has been a documented issue for a number of decades, especially in Mexico city (Castro, 2006; Quintero, 2004). This pork-barrel politics goes contrary to an ideal public participation in the management of water supply and sanitation services (WSS), which could explain the delay in the universal cover of WSS (Torregrosa-Armentia and Jimenez-Cisneros, 2009). Additionally, water utilities across the country have a high level of non-revenue water-ranging from 30 to even 50 % (SEMARNAT, 2009) and low price per cubic meter, which inhibits cost recovery for the water utilities that cannot invest in construction of new infrastructure and maintenance of the distribution network, an argument often used for warranting privatization, despite international evidence on the contrary (Schouten and Schwartz, 2006). Water prices remain politically contentious as politicians do not want to implement unpopular policies, such as an increment in the price of water, therefore water utilities remain helplessly dependent on subsidies to keep operating. In the year 2017 a cutback was announced of around 70 % to these subsidies, a situation that water advisors to the federal congress deem as potentially critical for water utilities across the country (García, 2016), which may start dropping the already subpar quality of WSS. Water utilities cannot guarantee the quality of tap water for human consumption in many cities of the country, triggering one of the most profitable bottled water markets in the world (Pacheco-Vega, 2015), with grim consequences for those 55 million poor, who spend a high percentage of their income on water, ranging from 5 to 15 % (Greene, 2017).

Aquifers are the main water source for large cities, providing up to 70 % of total water supply. This has led to a serious problem of land subsidence. Mexico City, for example, draws water from approximately 600 wells ($59\,m^3\,s^{-1}$), and land level is dropping at a rate of 12 to 22 cm yr^{-1} (Kimmelman, 2017). Mexico City only has 150 m^3 per capita per year of water availability, already counting several interbasin water transfer schemes. The Cutzamala scheme, one of the largest interbasin transfers in the world and usually praised for its engineering, is also the reason for an on-going water conflict with local indigenous people in the donor basin with severe water scarcity due to the water transfer (Gómez-Fuentes, 2014; Watts, 2015). Despite water conflicts, Conagua is already looking for new sources of water for another water transfer, as Mexico City is still running on deficit.

The tendency of Mexican cities is to keep growing while the countryside stagnates or even loses its population due to migration to cities. At the same time, rural needs have remained a point of contention for the lack of investment in infrastructure, which the urban areas quadruples that of rural areas (Conagua-Semarnat, 2015). This disparity in terms of investment in water infrastructure and other services, in which 3 out of 4 localities are considered marginalized (Conapo, 2017a), has led to an intense migration from rural to urban areas of people looking for a better living standards and opportunities.

1.4 Vulnerability under stochastic meteorological events – droughts and floods

In a climate change scenario provided by IPCC a reduction of 15.2 % rain is expected by 2030 and an increased pressure on water resources in the northern part of the country, while in the south more precipitation is expected and more floods (Martínez-Austria and Patiño-Gómez, 2012).

Mexico is prone to droughts, the most recent one of 2011–2012 being the worst in 70 years, with an agricultural economic loss of USD 1.3 billion, and an overall loss of 10 % of the GDP (Ortega-Gaucin et al., 2016). The country has shifted from a crisis management approach to a risk management one (Korenfeld-Felderman et al., 2014), through the National Drought Program (PRONACOSE for its Spanish acronym). The program is intended to lay out a set of preventive and mitigation actions to enhance the socio-ecological resilience should droughts strike. However, the measures planned to undertake in cases of drought, although a general improvement from the previous approach, whose sole measure was the access to a disaster fund and some relief measures, still is largely based on the status quo that elicits the other drivers of water crises.

Mexico City, Guadalajara, Monterrey and Leon, the largest cities in Mexico, present a recurrent problem on floods and water scarcity (Delgado-Ramos, 2015; Peña-Ramirez, 2012). These cities have undertaken simultaneously expensive infrastructural works at controlling urban floods and interbasin water transfers (Gobierno de Mexico, 2014). This situation can be explained by chaotic urbanization and land use, land speculation, and the loss of natural capital. However, instead of tackling these underlying causes, these cities persistently aim at solving these recurrent problems by developing large infrastructure and moving the water transfer frontier continuously further. This solution lacks flexibility and resilience to extreme meteorological events (Delgado-Ramos, 2015), and leaves the cities vulnerable to extremes such as droughts and floods, which especially affects the poor in peri-urban areas (von Bertrab and Wester, 2005).

2 Discussion

The combination of the four water crises described have at least three implications: the non-compliance with the human right on water and sanitation, ecosystems collapse and water conflicts. As the water stock is finite and diminishing faster than it is being replenished, and the population and the economy keeps growing, the gap between water supply and demand will continue to widen further. As cities most likely will keep accommodating the population and economic growth, they will be more dependent on water transfers and other large infrastructures to keep up with the demand for WSS, but always lagging behind the real demand. With scarce financial resources, water utilities in urban and rural settings will never reach a 100 % coverage in WSS, a situation that is advantageous for political careerism who address the water deficit partially and temporally in times of elections. Competition for water resources will become ever more acute as the agriculture, the largest water user, does not have the resources to modernize its infrastructure to diminish its wasteful use of water, and the higher returns for industry and public use shall translate in more pressure to reallocate water. Therefore, combined with a flawed water treatment system, the biggest loser will be the ecosystems and people dependent on them. Finally, the chaotic urbanization and the rising pressure in water resources and extreme meteorological events may increase the vulnerability of urban residents to floods and droughts, causing even more water conflicts.

3 Conclusions

This study shows that there are two kinds of drivers of water crises in Mexico: population and economic growth and water management and governance failures. Serious consideration and investment is only put into the first kind of water drivers through supply augmentation through large infrastructural schemes like dams, water transfers and water treatment plants. Although they may be deemed necessary for WWS, they also need to be complemented by institutional improvements that can anticipate and address water crises at a more fundamental level. Some policies that can improve this second kind of driver include the following:

- water utilities' independence from local and regional politics and funding,

- public participation in decision-making processes for water distribution agreements, in the design and scale of infrastructure, and in setting compensations for affected populations,
- transparency in contracts,
- access to reliable hydrological information,
- tackling pervasive corruption in allocation of water rights,
- updated regulation and norms of water pollution,
- law enforcement of water use and water pollution.

Although these polices are standard, the question lies in how to effectively implement them. The Lerma-Chapala case study in Mexico, is considered a success story regarding the management of a water crisis of a closed basin (Hidalgo and Pena, 2009). However, the IWRM approach used there have not yet been applied to other basins facing other kind of crises. As noted in Lenton and Muller (2009) international experiences teach that better water management is usually dependent on political processes outside the water sector. Political feasibility is key to policy changes based on different water management paradigms (Allen, 2003). Mexico is currently in the arduous political process of implementing a new fully coherent water policy based in tackling the drivers of water crises described.

Data availability. Data are available upon request by the corresponding author.

Competing interests. The authors declare that they have no conflict of interest.

Special issue statement. This article is part of the special issue "Water security and the food-water-energy nexus: drivers, responses and feedbacks at local to global scales". It is a result of the IAHS Scientific Assembly 2017, Port Elizabeth, South Africa, 10–14 July 2017.

Acknowledgements. I want to thank Heliodoro Ochoa-García for his commendable feedback and comments on this paper.

Edited by: Barry Croke
Reviewed by: two anonymous referees

References

Aboites, L., Cifuentes, E., Jimenez, B., and Torregrosa, L.: Agenda del agua, Mexico city: Academia Mexicana de Ciencias, 2008.

Allan, J. A.: IWRM/IWRAM: a new sanctioned discourse? School of Oriental and African Studies, Occasional, 1–27, 2003.

Castro, J. E.: Water, Power and Citizenship Social Struggle in the Basin of Mexico, New York: Palgrave Macmillan, 2006.

COMDA, MAPDER, AMCF, TODXS, A. P., DESC, E., RGEMA, MUPCND: Informe sobre violaciones a los Derechos Humanos al Agua y Saneamiento en México, 2017.

Conagua-Semarnat: Plan Nacional Hídrico 2013–2018, 2014.

Conagua-Semarnat: Situación del subsector agua potable, alcantarillado y saneamiento a diciembre de 2015, 2015.

Conagua-Semarnat: Estadísticas del Agua en México 2016, available at: www.gob.mx/conagua (last access: 7 September 2017), 2016.

Conapo: Localidades rurales, available at: http://www.conapo.gob.mx/es/CONAPO/Localidades_rurales (last access: 4 June 2017), 2017a.

Conapo: Proyecciones de población, available at: http://www.conapo.gob.mx/es/CONAPO/Proyecciones_Datos%0D, (last access: 25 May 2017), 2017b.

Coneval: Pobreza en México, available at: http://www.coneval.org.mx/Medicion/Paginas/PobrezaInicio.aspx, last access: 30 May 2017.

Delgado-Ramos, G. C.: Water and the political ecology of urban metabolism: the case of Mexico City, Journal of Political Ecology, 22, 98–114, 2015.

García, I.: Crisis líquida, Reporte Índigo, available at: http://www.reporteindigo.com/reporte/mexico/pef-escazes-agua (last access: 7 September 2017), 2016.

Gobierno de Mexico: Programa Nacional de Infraestructura 2014–2018, available at: http://www.dof.gob.mx (last access: 7 September 2017), 2014.

Gobierno de Mexico: Reformas estructurales y expectativas de crecimiento de la economía y de la productividad, available at: https://www.gob.mx/cms/uploads/attachment/file/160126/II_Reformas_Estructurales-1.pdf (last access: 7 September 2017), 2017.

Gómez-Fuentes, A. C.: Estado y política hidráulica en México: el conflicto de los indígenas mazahuas, Agua Y Territorio, 4, 21–29, 2014.

Gómez-Valdez, M. I. and Palerm-Viqueira, J.: Abastecimiento de agua potable por pipas en el valle de texcoco, méxico, Agricultura, Sociedad Y Desarrollo, 12, 567–586, 2015.

Greene, J.: Peri-urban bottled water consumption in western Mexico, Unpublished raw data, 2017.

Hansen, A. M. and Juárez-Corzo, C.: Evaluation of the Pollution of Hydrological River Basins: Priorities and Needs, in: Water Resources in Mexico: Scarcity, Degradation, Stress, Conflicts, Management, and Policy, edited by: Spring, U. O., Berlin, Heidelberg: Springer Berlin Heidelberg, 201 pp., 2011.

Hidalgo J. and Pena, H.: Turning water stress into water management success: experiences in the Lerma-Chapala River Basin, in: Integrated Water Resources Management in Practice, edited by: Lenton, R. and Muller, M., Earthscan: London, 107–120, 2009.

INEGI: Encuesta intercensal 2015, available at: http://www3.inegi.org.mx/sistemas/temas/default.aspx?s=est&c=17484 (last access: September 2017), 2015.

Kimmelman, M.: Mexico City, Parched and Sinking, Faces a Water Crisis, New York Times, available at: https://www.nytimes.com/interactive/2017/02/17/world/americas/mexico-city-sinking.html?action=clickandcontentCollection=Opinionandmodule=Trendingandversion=Fullandregion=Marginaliaandpgtype=articleand_r=0, last access: 7 September 2017.

Korenfeld-Federman, D., Arreguín, F. I., and López-Pérez, M.: Constructing a framework for National Drought Policy: The way forward in Mexico, Weather and Climate Extremes, 3, 90–94, https://doi.org/10.1016/j.wace.2014.04.003, 2014.

Lenton, R. and Muller, M.: Integrated water resources management in practice: Better water management for development, Routledge, 2009.

Lezama, C.: Construcción social de ciudades no sustentables: el caso de El Salto, Jalisco, in: The Politics of Deterioration: The Urban-Industrial Dynamics of the Santiago River, Jalisco, Mexico, edited by: McCulligh, C., Santana, L., and Lezama, C., Newcastle: WATERLAT-GOBACIT NETWORK WORKING PAPERS, 17–36, 2016.

Madani, K.: Water management in Iran: what is causing the looming crisis?, Journal of Environmental Studies and Sciences, 4, 315–328, https://doi.org/10.1007/s13412-014-0182-z, 2014.

Martínez-Austria, P. F. and Patiño-Gómez, C.: Efectos del cambio climático en la disponibilidad de agua en México, Tecnologia Y Ciencias Del Agua, 1, 5–20, 2012.

McCulligh, C.: Alcantarilla del progreso: industria y Estado en la contaminación del río Santiago en Jalisco, CIESAS, 2017.

Mollard, E., Vargas, S., and Wester, P.: Social Participation in Mexican River Basin Organizations: The Resilience of Coalitions, edited by: Berry, K. A. and Mollard, E., Social Participation in Water Governance and Management, 115–135, Earthscan, 2010.

Ortega-Gaucin, D., López-Pérez, M., and Arreguín-Cortés, F. I.: Drought risk management in Mexico: progress and challenges, in: 10th International Conference on Risk Analysis, 6, 161–170, https://doi.org/10.2495/SAFE-V6-N2-161-170, 2016.

Pacheco-Vega, R.: Agua embotellada en México: de la privatización del suministro a la mercantilización de los recursos hídricos, Espiral, 63, 221–263, 2015.

Peña-Ramirez, J.: Crisis del agua en Monterrey, Guadalajara, San Luis Potosí, León y la Ciudad de México, Mexico, D.F., Universidad Nacional Autónoma de México, 2012.

Quintero, S. G.: El uso político del agua, Algunos ejemplos y reflexiones, edited by: Ramirez, J. P., El agua, espejo de los pueblos Mexico, D.F.: Plaza y Valdés, 265–281, 2004.

Ramón-Lahera, V.: Infraestructura sustentable: Las plantas de tratamiento de aguas residuales, Quivera, 12, 58–69, 2010.

Richter, B.: Chasing Water, Island Press, https://doi.org/10.5822/978-1-61091-537-3, 2014.

Schouten, M. and Schwartz, K.: Water as a Political Good: Implications for Investments, International Environmental Agreements, 6, 407–421, https://doi.org/10.1007/s10784-006-9013-3, 2006.

Scott, C. A. and Banister, J. M.: The Dilemma of Water Management "Regionalization" in Mexico under Centralized Resource Allocation, International Journal of Water Resources Development, 24, 61–74, 2008.

SEMARNAT: Agua producida, facturada y no contabilizada en ciudades seleccionadas con más de 500 mil habitantes, available at: http://apps1.semarnat.gob.mx/dgeia/compendio_2013/archivos/01_agua/D3_AGUA07_07.pdf (last access: 7 September 2017), 2013.

Stoltenborg, D. and Boelens, R.: Disputes over land and water rights in gold mining: the case of Cerro de San Pedro, Mexico, Water Int., https://doi.org/10.1080/02508060.2016.1143202, in press, 2016.

Torregrosa-Armentia, M. L. and Jimenez-Cisneros, B.: Challenges Facing the Universal Access of Water and Sanitation in Mexico, edited by: Castro, J. E. and Heller, L., Water and Sanitation Services, Earthscan, 338–347, 2009.

von Bertrab, É. and Wester, P.: Gobernabilidad del agua en México, La crisis de agua de Guadalajara y el destino del Lago de Chapala, Medio Ambiente Y Urbanización, 62, 143–160, 2005.

Watts, J.: Mexico City's water crisis – from source to sewer, The Guardian, available at: https://www.theguardian.com/cities/2015/nov/12/mexico-city-water-crisis-source-sewer (last access: 7 September 2017), 2015.

Wester, P., Hoogesteger, J., and Vincent, L.: Local IWRM organizations for groundwater regulation: The experiences of the Aquifer Management Councils (COTAS) in Guanajuato, Mexico, Natural Resources Forum, 33, 29–38, 2009.

World Economic Forum: Water Security: The Water-Food-Energy-Climate Nexus, Island Press, Washington, D.C., 2011.

Proc. IAHS, 376, 63–68, 2018
https://doi.org/10.5194/piahs-376-63-2018

Conceptual framework to ensure water security in Ukraine

Yaroslav Gadzalo[1], **Mykhailo Romashchenko**[2], **and Mykhailo Yatsiuk**[2]

[1]National Academy of Agrarian Sciences of Ukraine, Kyiv city, 01010, Ukraine
[2]Institute of Water Problems and Land Reclamation of NAAS, Kyiv city, 03022, Ukraine

Correspondence: Mykhailo Romashchenko (mykhailo.rom@gmail.com)

Received: 7 June 2017 – Accepted: 28 September 2017 – Published: 1 February 2018

Abstract. As a result of global climate change against the background of natural water supply deterioration and river water content reductions, nowadays Ukraine is facing the problem of environmental degradation of river basins.

In light of this, we suggest that achieving an acceptable level of water security in Ukraine should be defined as the strategic objective of national water policy. The state of national water security should be evaluated by its progress in certain sectors. The basic principles of the new water policy of Ukraine are supposed to be represented in Water Strategy of Ukraine. Integrated water management by the basin principle should serve as the main tool for achieving the objectives of water security.

According to global estimates the amount of suitable water for human consumption and functioning of most terrestrial ecosystems on the planet is 35 million km^3 that is only 2.5 % of the total Earth water resources, of which two-thirds is contained in glaciers and snow cover, making them virtually inaccessible for widespread use (FAO, 2011).

The scarcity of available fresh water resources and their uneven spatial distribution over the world, as well as the increase of water consumption, water pollution and climate change are gradually transformed into global problems related to drinking water shortage, from the one side, and threats of regional water conflicts, from the other side.

That is the reason why, starting from the beginning of the XXI century, a concept of "water security" has been coming into widespread acceptance and is a cornerstone of the modern vision of the future of our planet. Water security can be defined as "the availability of a suitable quantity and quality of water for health, life activities, ecosystems and production, coupled with an acceptable level of water-related risks to people, environment and economy" (Grey and Sadoff, 2007). So, achieving an appropriate level of water security is not only an ultimate objective of water policies to growing number of countries, but also one of the main goals in sustainable development of mankind in general.

As defined by the International Water Management Institute (IWMI, 2014) the concept of water security is an integral part of the agenda to address global problems. The concept recognizes the importance of synergy between water-dependent sectors and the need to strengthen these links in order to create an effective water management and establish appropriate policies.

Global Water Partnership (GWP, 2014) declares that the concept of water security is based on the recognition of the key role of water to all areas of human development: food security, health, poverty reduction and sustainable development in agriculture, industry and energy sectors.

The state of water resources and water supply for population and industries of Ukraine remains a major threat to the national security. The severity of this problem arises from the fact that Ukraine is one of the least sustainable countries in Europe regarding its own water resources having the intensity of local runoff in dry years per capita only 0.6 thousand m^3, and including the transit flow – 1.2 thousand m^3 (Romashchenko et al., 2015) (Fig. 1).

The average long-term runoff in Ukraine is 54.7 km^3 a year, and it varies in different years by dryness – 50.9 km^3 – in average year, 38.8 km^3 – in dry years and 25.6 km^3 – in very dry years (Table 1).

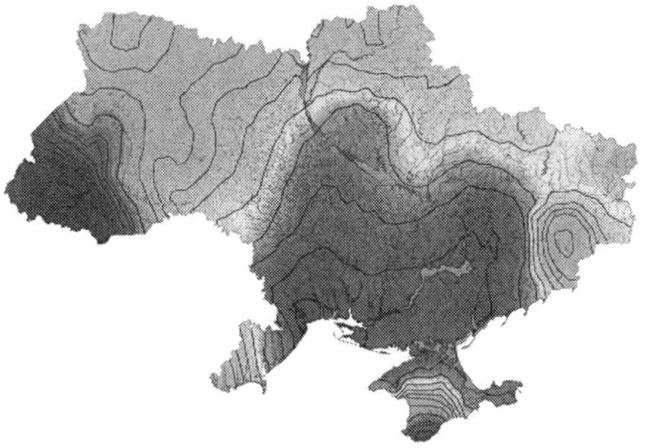

Figure 1. Distribution of the average rate of runoff over the territory of Ukraine.

Table 1. Water resources of Ukraine.

Resources	Water resources by year by water content km^3	
	Medium	Low water
Inflow of transit river runoff	157.4[a]/37.3[b]	121.7[a]/26.2
Local river runoff	52.4	29.7
Total resources of river runoff	209.8/87.1	151.4/55.9
Predicted resources of ground waters	22.5	22.5
Including those, which are not involved in the surface runoff	7.0	7.0
Total resources of fresh water	216.8/94.1	158.4/62.9

[a] including 122.7 and 95.5 km^3 along the Kiliyske river mouth of the Danube river; [b] excluding the Danube river.

The difference in the availability of local water resources between some regions of the country is very significant, and can be 60 times as much from 0.14 km^3 a year in Kherson region against 7.92 km^3 a year – in Transcarpathian region.

The surface water available for use is very unevenly allocated across the country. More than half of the water is concentrated in the Danube basin, where water demand does not exceed 5 %. To improve the water supply in water-stressed regions more than 1160 reservoirs with a total storage capacity of about 55 km^3 have been constructed in Ukraine. The network of supply canals (over 1.0 thousand km) and water pipelines (over 2.0 thousand km) is also well-developed. This enables to redistribute respectively 3 and 12 km^3 of water annually through the territory of the country (Romashchenko et al., 2015).

River regularization of the main watercourses of Ukraine made it possible to redistribute the runoff and to meet the needs of water consumers and water users, but turned the rivers into significantly altered water bodies with poor ecological status and low potential of self-purification. It is demonstrated by the lack of any significant improvement in the quality of water in the Dnieper or other rivers despite an intense reduction in water intake (more than three fold for the last 25 years: from 30 km^3 in 1990 up to <9.7 km^3 in 2015) and the discharges of sewage waters (from 18 km^3 in the 1990s to 5.3 km^3 in 2015) in different sectors of the country's economy. The runoff regulation is accompanied by the emergence of a number of environmental problems related to the functioning of water bodies. In particular, the well-known consequences of the construction of hydroelectric power stations are the creation of hydrotechnical structures of varying complexity and purpose, the operation of which gives rise to the hydromorphological changes in the rivers (the loss of flow speed, the rise of water levels, the accumulation of significant volumes of bottom sediments, flooding of large areas for reservoir beds formation and so on) that in turn causes the degradation of river ecosystems and the loss of their ability to self-purification as well as the significant pollution of surface waters. That also affects the qualitative and quantitative state of biological resources and leads to the flooding of adjacent areas.

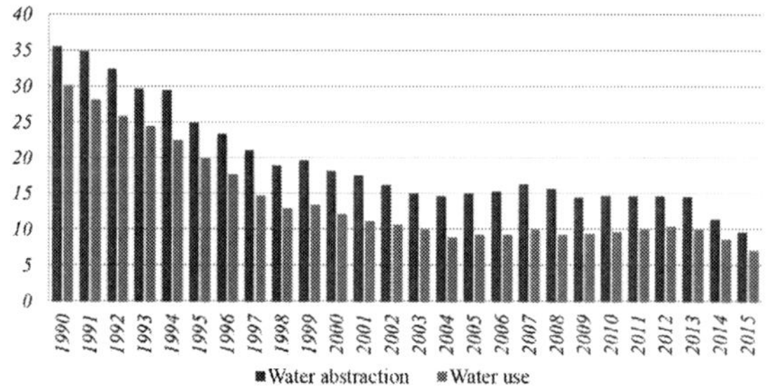

Figure 2. Water abstraction and water use in Ukraine, billion m^3.

Following a long (more than 30 years) tendency to warming and rising the frequency and duration of droughts (especially for the last 10 years) as well as to the reduction in river flow, the issue of regulating the operation regimes of reservoirs and distributing the accumulated water to provide the population and economy with water in accordance with the approved priorities is becoming more and more a difficult task. In the recent years, characterized by low water content, the water accumulated in the reservoirs is not only enough for environmental but even for regulating releases.

Annually in Ukraine about 11 km^3 of water are extracted for the needs of water supply and economic sectors, including about 8 km^3 – from surface sources, 2 km^3 – from underground sources and up to 1 km^3 – sea water.

Under the climate change some additional threats to food security are emerging due to the rise of the deficit of annual water balance in a large part of the country and deterioration of the engineering infrastructure of drainage systems as well as the stagnation of the utilization of irrigation and drainage potential (Gadzalo et al., 2014). Utilizing the potential of irrigation and drainage systems as a basis for improving natural water supply is unsatisfactory, since from the existing

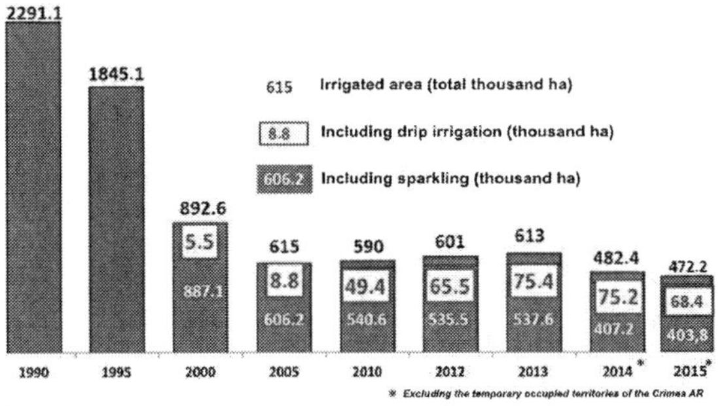

Figure 3. Actual irrigated area.

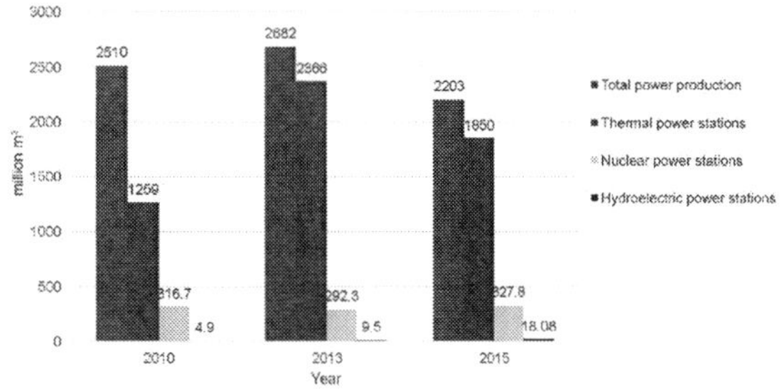

Figure 4. Total water abstraction from natural water bodies for the production of electricity, million m^3.

2.17 million hectares of irrigated land only 500 thousand are actually irrigated while the bilateral water control is exercised only on 250 thousand from the available 3.2 million hectares of drained land. That resulted in the disability of reclaimed land to act as a security reserve in food and resources' provision of the state in the years of adverse weather conditions. Therefore the task of restoring an effective environmentally sustainable utilization of reclaimed land potential, especially irrigation, is one of the priorities of the state policy of Ukraine as well as one of key aspects of the development of the agricultural sector of the economy. Specifically the development of irrigation and drainage sector is considered to be as an objective prerequisite for achieving the annual grain production in the quantity of 80–100 million tons that will enable to turn Ukraine into one of the world food donors.

As concerns the issue of water supply of cities and households the main problem is the insufficient level of centralized water supply and sanitation, since more than 1300 villages with the populations over 800 thousand people so far use imported water (Adamenko et al., 2016) (Table 2).

Traditional energy production in Ukraine (at TPP or NPP) is extremely water-intensive process. In 2015, the sector of thermal power used for its own needs 2400 million m^3 of water, the sector of nuclear power – 327.8 million m^3, 106 000 million m^3 of water ran through the turbines of HPP (the runoff of the Dnieper River – the main waterway of Ukraine in a dry year is 30 000 million m^3); but only about 11 billion kWh were produced, which indicates the inefficient use of water resources in the electricity generation industry.

Thermal and nuclear power plants that require large amounts of water for electricity generation and cooling turbines and reactors pose significant risks to water security. The operation of direct-cooling systems of thermal power plants, which requires the significant amounts of water and leads to a decrease in the runoff of water bodies and deterioration of water quality (thermal pollution), is causing concern. In dry (low water) years, the operation of thermal and nuclear power plants greatly steps up the "competition" for water resources in Ukraine. Due to the limited use of water resources, the issue of potential development and completion with new blocks of existing NPPs deserves special attention. Thus, in 2007, under the conditions of drought, the issue of limiting access to water for drinking needs and agriculture in the basin of the Southern Bug was acutely due to the complete regulation of the runoff in the Oleksandrivske reservoir to ensure the safe operation of the South-Ukrainian NPP.

In aquatic ecosystems sector it can be observed a general trend to the deterioration of water quality in river basin, in other words – metamorphization of chemical water composition (from calcium bicarbonate water to sulphate sodium chloride water). The load on the aquatic ecosystems of river basins (the ratio of water intake in the basin to the volume of actual runoff) in most basins is above the environmentally safe level and in the basin of the Siversky Donets River it is critical and accounts for 42 % (Romashchenko et al., 2015) (Table 3).

The modern system of river runoff and water quality monitoring requires to increase a number of monitoring posts and update their and laboratory equipment, computerize the data collection and information management service; develop new mechanisms for accessing the information for the rapid adoption of management decisions.

The problems related to protection against the harmful effects of water have not solved yet to the full extend. The third of the country's population regularly suffers from floods and waterlogging. The rivers of the Carpathian region are characterized by the increased flood hazard. About 11 % of the territory of Ukraine (over 540 cities and towns, more than 450 villages) are prone to subsoil waterlogging.

Despite a large number of water and ecological problems that really affect the national security of Ukraine, the problem of establishing a proper state of water security in Ukraine at the legislative level as an important component of national security has not been resolved yet. It till requires further scientific rationale and regulatory support as a strategic goal of the country's sustainable development. Ukraine has not developed and adopted the water strategy for Ukraine yet, and

Table 2. Centralized water supply of settlements in Ukraine.

Settlements	Population, '000 people	Years						
		2000	2009	2010	2011	2002	2013	2014
Towns	More than 500	93	91.3	991	99.4	99.6	99.8	89.8
	100–500	86						
	20–100	75						
Small towns	20		87.1	87.1	87.0	85.9	85.9	68.8
Rural settlements	Less than 20	26	22.1	22.0	22.0	22.1	22.1	17.2

Table 3. Water abstraction from the river basins of Ukraine.

No.	River basin	Catchment area, km^2	Natural runoff, km^3	Actual runoff, km^3	Abstraction	
					km^3	%
1	Dnieper	540 000	53.5	41.9	8.68	20.7
2	Dniester	72 100	10.0	8.66	0.62	7.1
3	Siversky Donets	52 400	3.69	3.47	1.45	41.8
4	Southern Bug	63 700	3.02	2.83	0.39	13.8

subsequently the strategic goals of the state water policy and the objectives to achieve it have not been defined and formed either for a long- or near-term prospect.

The first positive step towards solving this problem was holding the National Water Dialogue in March 2017 at which the issue of the necessity of water strategy development for Ukraine as a basis for the formation of water policy of the country was raised (Adamenko et al., 2016).

According to the proposed evaluation system (ADB, 2013) the level of water security of Ukraine is defined as threatening due to a partial legislative support in the field of water management and environmental protection, insufficient public investment and failure to abide by the rules and laws in this sector.

In light of this, achieving an acceptable level of water security in Ukraine is proposed to determine as a strategic objective of national water policy. Thus the national targets and indicators of water safety should be specified in view of the obligations undertaken by Ukraine under the Association Agreement between Ukraine and the European Union, the provisions of the EU Water Framework Directive 2000/60/EC and national targeted indicates to the Protocol on Water and Health. The strategic goals and objectives of the national environmental policy, the objectives of the "State target program of water management and environmental rehabilitation of the Dnieper River basin for the period until 2021 year" as well as the provisions of the recently adopted Law of Ukraine "On amendments to some legislative acts of Ukraine concerning the implementation of integrated approaches to water management by basin principle" should be also taken into account for this purpose. Water Strategy of Ukraine should become the main legal document that defines the principles of the new water policy of Ukraine.

Integrated water management by the basin principle should be the main instrument of timely and proper fulfilling the tasks to achieve a good level of water security, subject to an adequate scientific support for the conceptual and applied aspects of the management (Adamenko et al., 2016).

Despite the significant scientific achievements on the integrated water management, its implementation in practice just as it is will not allow to achieve a proper level of water security in Ukraine without carrying out a complex of additional research.

To form the scientific principles of water security of Ukraine in terms of climate change the Presidium of the National Academy of Agrarian Sciences of Ukraine proposed the Presidiums of the National Academy of Sciences and National Academy of Medical Sciences to co-develop the concept of the scientific and technical program for 2018–2025 years "Scientific principles of ensuring water security of Ukraine", the main areas of which should include:

- developing scientific and methodological, organizational and legal as well as legislative principles of the formation of water security of the country in terms of climate change and growing water stress;
- developing scientific and methodological foundations to improve the criteria and indicators of national and sectorial water securities, establishing and performing their monitoring for effective water management;
- studying the water cycle processes and developing environmentally-friendly and economically reasonable methods of water extraction from the natural hydrologic cycle to meet the needs of social development;

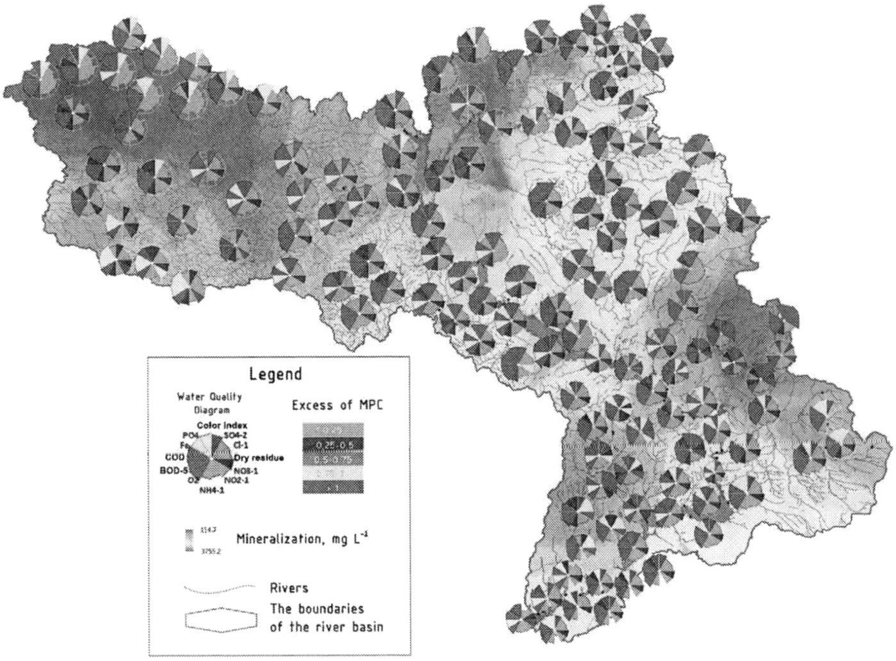

Figure 5. Map of indicators of surface water quality in the Dnieper River basin.

- developing water use technology that provides a decrease in specific water consumption per unit of goods produced in all sectors of the economy.

In general, for today the issue of the development of legislative, organizational, legal, scientific and methodological bases for the formation of water security of the country in terms of climate change and its impact on environmental and economic security of the state, becomes a key aspect and requires to be one of the national security issues followed by the further refining of scientifically grounded management decisions at the national level. Therefore, the Presidium of the National Academy of Agrarian Sciences of Ukraine appealed to the National Security and Defense Council of Ukraine with the proposal to consider at a meeting of NSDC the question "On the state and objectives for improving water security in Ukraine" as one of the priority issues of national security of Ukraine.

In addition, it has been proposed to put in the list of the main goals of the "Sustainable Development Strategy of Ukraine" the task of achieving a good level of water security in Ukraine up to 2050.

In addition, it was proposed to include in the list of the main goals of the sustainable development of the Strategy of Sustainable Development of Ukraine the task of achieving a good state of water security in Ukraine up to 2050.

It can be possible subject to a set of measures to be implemented, the main of which are:

- to harmonize and implement water legislation in accordance with the Association Agreement between Ukraine and the EU, in particular, to implement the integrated river basin management, which is the way of providing water security;
- to review, optimize and update the water resources monitoring system (in accordance with the requirements of the EU Water Framework Directive);
- to state the provision of water and hygiene as a priority number 1 in the allocation of available water resources among the various water users and develop the necessary mechanisms (including economic ones) to implement this priority in practice;
- to develop and implement the Plan of Implementation of the Protocol on Water and Health, taking into account national targets and indicators;
- to approve and implement methodological recommendations for the preparation of River Basin Management Plans (RBMPs), stimulate the development and implementation of RBMPs taking into account both flood and draught risk managements that are relevant for the whole of Ukraine;
- to review the state target water programs, in particular, "Drinking Water of Ukraine", "Program for the De-

velopment of Water Management and Ecological Improvement of the Dnieper River Basin for the period up to 2021"), identify the priority areas and measures and ensure an appropriate public funding using the co-financing mechanism for interested parties.

Data availability. The main research data were used from various storage facilities – the Central Geophysical Observatory (for the construction of Fig. 1), the statistics of the State Agency of Water Recourses of Ukraine (for the construction of Figs. 2, 3, 4), the Dnieper River Basin Department of the State Agency of Water Recourses of Ukraine (for the construction of Fig. 5, http://dbuwr.com.ua/upravlinnya-vodnimi-resursami/monitoring-yakosti-vodi.html, Sreznevsky, 2015).

Competing interests. The authors declare that they have no conflict of interest.

Special issue statement. This article is part of the special issue "Water security and the food–water–energy nexus: drivers, responses and feedbacks at local to global scales". It is a result of the IAHS Scientific Assembly 2017, Port Elizabeth, South Africa, 10–14 July 2017.

Edited by: Barry Croke
Reviewed by: two anonymous referees

References

Adamenko, T. I., Demydenko, A. O., Romashchenko, M. I., Tsvietkova, A. M., Shevchenko, A. M., and Yatsyuk, M. V.: Rethinking of Water Security for Ukraine, 19, Kyiv, available at: http://www.gwp.org/globalassets/global/gwp-cee_files/regional/rethinking-water-security-ukraine-2016.pdf (last access: 8 December 2017), 2016.

ADB (Asian Development Bank): Asian Water Development Outlook 2013: Measuring water security in Asia and the Pacific, Mandaluyong City, Philippines, Asian Development Bank, available at: https://www.adb.org/sites/default/files/publication/30190/asian-water-development-outlook-2013.pdf (last access: 8 December 2017), 2013.

FAO: The state of the world's land and water resources for food and agriculture (SOLAW) – Managing systems at risk. Food and Agriculture organization of the United Nations, Rome and Earthscan, London, 2011.

Gadzalo, Y. M., Romashchenko, M. I., and Gryn, Y. I.: Concept of restoration and development irrigation in Southern region of Ukraine, 27, Kyiv, 2014.

Grey, D. and Sadoff, C. W.: Sink or Swim? Water security for growth and development, Water Policy, Official Journal of the World Water Council, 9, 545–570, 2007.

GWP: Strategy Towards 2020: A water secure world, 28, Stockholm, available at: http://www.gwp.org/globalassets/global/about-gwp/strategic-documents/gwp_strategy_towards_2020.pdf (last access: 8 December 2017), 2014.

IWMI: Strategy 2014–2018, Solutions for a water-secure world, 32, Sri Lanka, available at: http://www.iwmi.cgiar.org/About_IWMI/PDF/iwmi-strategy-2014-2018.pdf (last access: 8 December 2017), 2014.

Romashchenko, M. I., Hvesyk, M. A., and Mykhaylov, Y. O.: Water Strategy of Ukrainefor the period until 2025 year, Scientific foundations, 46 pp., 2015.

Sreznevsky, B.: Central Geophysical Observatory, available at: http://dbuwr.com.ua/upravlinnya-vodnimi-resursami/monitoring-yakosti-vodi.html (last access: 8 December 2017), 2015.

Proc. IAHS, 376, 69–75, 2018
https://doi.org/10.5194/piahs-376-69-2018

Open Access PIAHS

Performance assessment of the Gash Delta Spate Irrigation System, Sudan

Araya Z. Ghebreamlak[1], Haruya Tanakamaru[1], Akio Tada[1], Bashir M. Ahmed Adam[2], and Khalid A. E. Elamin[2]

[1]Graduate School of Agricultural Science, Kobe University, Nada-ku, Kobe 657-8501, Japan
[2]Agricultural Research Corporation, Wad Medani, P.O. Box 126, Sudan

Correspondence: Araya Z. Ghebreamlak (araiag4@gmail.com)

Received: 6 June 2017 – Accepted: 22 August 2017 – Published: 1 February 2018

Abstract. The Gash Delta Spate Irrigation System (GDSIS), located in eastern Sudan with a net command area of 100 000 ha (an area currently equipped with irrigation structures), was established in 1924. The land is irrigated every 3 years (3-year rotation) or every 2 years (2-year rotation) so that about 33 000 or 50 000 ha respectively can be cultivated annually. This study deals with assessing the performance of the 3- and 2-year rotation systems using the Monte Carlo simulation. Reliability, which is a measure of how frequently the irrigation water supply satisfies the demand, and vulnerability, which is a measure of the magnitude of failure, were selected as the performance criteria. Combinations of five levels of intake ratio and five levels of irrigation efficiency for the irrigation water supply of each rotation system were analysed. Historical annual flow data of the Gash River for 107 years were fit to several frequency distributions. The Weibull distribution was the best on the basis of the Akaike information criteria and was used for simulating the ensembles of annual river flow. The reliabilities and vulnerabilities of both rotation systems were evaluated at typical values of intake ratio and irrigation efficiency. The results show that (i) the 3-year rotation is more reliable in water supply than the 2-year rotation, (ii) the vulnerability of the 3-year rotation is lower than that of the 2-year rotation and (iii) therefore the 3-year rotation is preferable in the GDSIS. The sensitivities of reliability and vulnerability to changes in intake ratio and irrigation efficiency were also examined.

1 Introduction

Spate irrigation is a type of surface irrigation in which floodwater from a seasonal river is diverted to a deep-soiled flat field and allowed to spread naturally under the influence of slope, soil type and surface roughness. It is an inefficient method of irrigation with little control of the water spreading over the field (James, 1988). Nonetheless, it is feasible when the following requirements are met (Zimmerman, 1966): (i) a large and flat area within the proximity of an intermittent stream is available; (ii) the soil of the area is deep with good permeability and high water-holding capacity; (iii) and a shallow stream requiring a low diversion structure exists near the area. Moreover, in arid and semi-arid environments, in addition to the difficulty of finding a suitable site for a reservoir, the classical approach to water management through water storage is not practical as very high sediments and evaporation loss would not justify the development of economically useful reservoirs (van Steenbergen et al., 2010). That is why spate irrigation is common in arid and semi-arid regions, typically in countries in the Middle East, North and East Africa, West Asia and parts of Latin America (van Steenbergen, 1997).

The Gash Delta Spate Irrigation System, hereafter abbreviated as the GDSIS, in Kassala state, eastern Sudan (Fig. 1) was established in 1924. Spate water from the Gash River with an average effective flood period of about 70 days is diverted through seven intake structures. As shown in Fig. 2, a network of seven main canals and 13 sub-canals conveys the diverted water to six irrigation blocks within a net command area of 100 000 ha (Cole, 1989). The flow in these gravity canals is divided into distributary canals that directly supply agricultural fields through breaches in the canal banks from

Published by Copernicus Publications on behalf of the International Association of Hydrological Sciences.

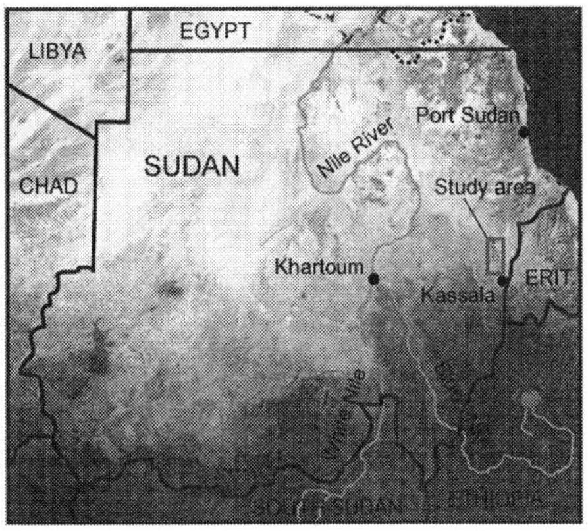

Figure 1. Location of the study area (source of satellite image: http://www.maplibrary.org/).

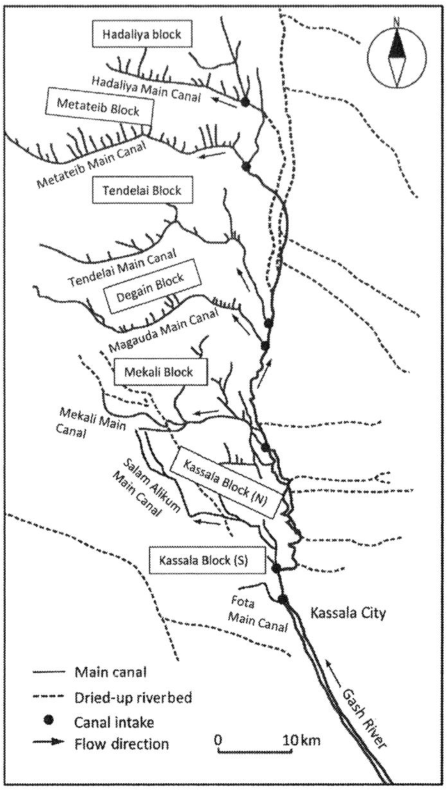

Figure 2. Schematics of the Gash Delta Spate Irrigation System (the original figure by Kirkby, 2001, was modified).

which it spreads over the land with minimum further control (Swan, 1959).

Under the original plan, the land is irrigated every 3 years (3-year rotation) so that about 33 000 ha of the 100 000 ha can be cultivated annually. Each tenant previously cultivated 4.2 ha of cotton as a cash crop and 0.42 ha of sorghum for food self-sufficiency. The system functioned well when the number of tenants registered was about 10 000. Starting from the 1970s, the performance went into serious decline and was paralleled by droughts and civil unrest. Kassala state was settled with about 20 % of the total refugees in Sudan, and as a result the number of GDSIS beneficiaries has increased fourfold. During the same period the cultivated area decreased by 50 % owing to deterioration of the irrigation infrastructure (IFAD, 2004). Consequently, the land tenancy entitlement for each farmer is reduced considerably and farmers are allocated far less irrigated area than their entitlements. Under such conditions, the International Fund for Agricultural Development (IFAD) supported a programme for rehabilitation of the irrigation diversion structures in 2003. As a result, some changes were made, such as the change from a 3- to a 2-year rotation system to accommodate more farmers. However, according to the evaluation reports by Anderson (2011), IFAD (2012) and IFAD (2014), the designed annual target irrigated area of the 2-year rotation (50 000 ha) has never been realistic. Owing to a lack of detailed scientific studies and the complexity of the system, the above evaluation reports have not concretely articulated the reasons for the failure.

Despite the variability in the annual flow of the Gash River, the target irrigated areas of the 3- and 2-year rotations were decided on the basis of the average flow. This study investigates the irrigation water supply performance of the GDSIS through a Monte Carlo simulation by considering the stochastic nature of the Gash River flow. Moreover, the irrigation water supply, which determines the size of the irrigated area, depends on the Gash River water allocation policy and capacity as well as the operation and efficiency of the irrigation structures. Hence, this study also deals with a parametric analysis of the effects of water allocation policy, planning and operation of the irrigation structures on the performance of the GDSIS. Towards this purpose, the proposed approach can be used to analyse different strategies for rehabilitation planning and operation policies and to select the best possible option.

2 Data and simulation model

2.1 Annual flow data and statistics

Analysis of the annual flow volume of the Gash River for 107 years (1907–2013) at Kassala Bridge, as shown in Fig. 3, was carried out to check for randomness and select the best fit probability density function.

The Gash River is a seasonal river, carrying water only during the rainy season from July to September. It is dry during the remaining months of the year. There is no flow carry-over from one year to the next. Moreover, a large part

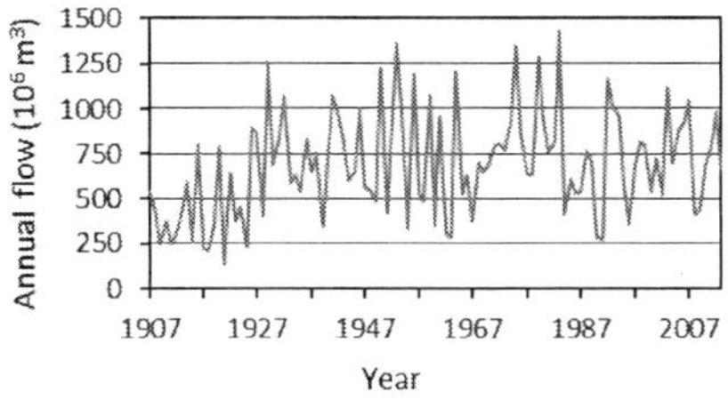

Figure 3. Annual flow of the Gash River measured at the Kassala Bridge gauging station (1907–2013; source: Gash Agricultural Scheme administrative office and Gash River training unit, Kassala).

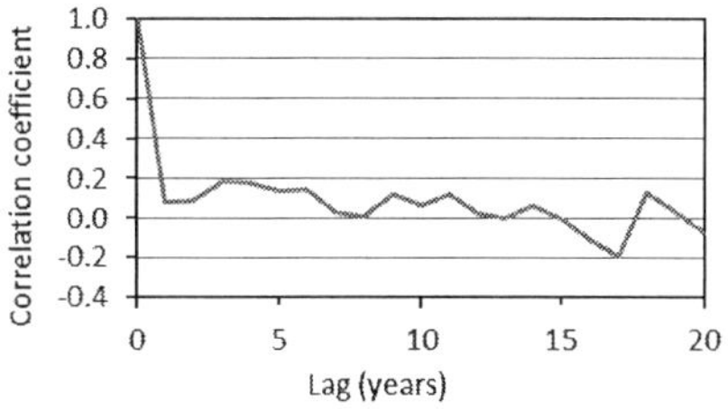

Figure 4. Correlogram of annual flow volume at Kassala Bridge (1907–2013).

of the river basin is in the semi-arid region. Thus annual flows in this river can be assumed as independent time series. The correlogram as depicted in Fig. 4 supports the above assumption.

The probability density functions commonly used in hydrology and water resources fall into three families. The first includes the normal and log-normal distributions. The second is the gamma and related distributions that include the exponential, log Pearson type III and beta distributions. The third family includes the extreme value distributions; special cases are the Gumbel and the Weibull distributions. In this study, normal, two- and three-parameter log-normal, two- and three-parameter gamma and two-parameter Weibull are tested.

Maximum likelihood method was used to estimate the parameter of each candidate. The chi-square test was used to check the goodness of fit at the 95 % confidence interval, and all the tested distributions were found to be adequate. The Akaike information criteria (AIC) was used to rank the selected probability distributions. The two-parameter Weibull distribution was found to be the best. This superiority of the Weibull over the others is also indicated in the Q–Q plot of Fig. 5.

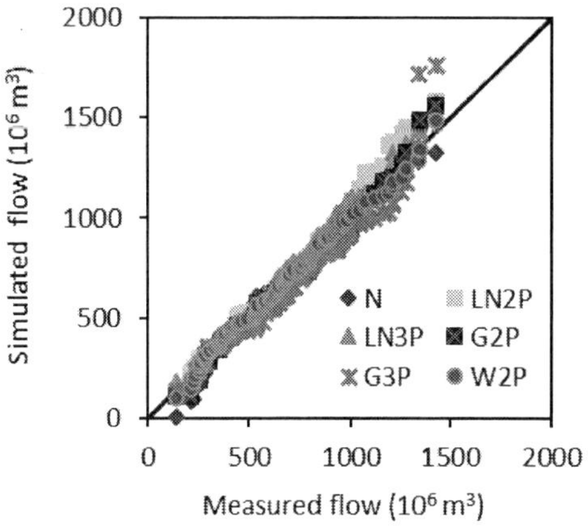

Figure 5. Q–Q plot of the selected probability density functions (N = normal, LN2P = two-parameter log-normal, LN3P = three-parameter log-normal, G2P = two-parameter gamma, G3P = three-parameter gamma and W2P = two-parameter Weibull).

2.2 Simulation model

There are many ways of summarising time series performance data that might result from simulation analysis. The simple ways frequently used are the mean and variance (Hashimoto et al., 1982; Loucks and van Beek, 2005). However, the mean and variance usually fail to show how severe and how frequent periods of poor performance may be. Thus, other criteria, such as reliability and vulnerability, have gained popularity in water resources studies (Asefa et al., 2014; Jain, 2010; Jain and Bhunya, 2008; Kjeldsen and Rosbjerg, 2004), and these criteria are used in this study.

The reliability (α) is defined as the frequency or probability that a system (X) is in a satisfactory state (Sat; Hashimoto et al., 1982). Therefore, the reliability of time series data can be defined as the amount of data in a satisfactory state (S) divided by the total amount of data (N) in the time series:

$$\alpha(\%) = \mathrm{Prob}\{X \in \mathrm{Sat}\} = \frac{S}{N} \times 100. \quad (1)$$

The vulnerability refers to the magnitude of a failure (here it is deficit, D) if one occurs. The expected value of vulnerability υ is given by Loucks and van Beek (2005):

$$\upsilon = \frac{1}{N-S}\sum_{i=1}^{N-S} D_i. \quad (2)$$

The mathematical descriptions of reliability and vulnerability are based on the assumption that the system under consideration at a given time is in either a satisfactory state or an unsatisfactory state. The focus of the current study is on an irrigation system. Hence, the satisfactory state occurs when the irrigation water supply is able to meet the irrigation water demand, and the unsatisfactory state occurs when the irrigation water supply is unable to meet the irrigation water demand.

For convenience of comparing the 3- and 2-year rotation systems, the original vulnerability of Eq. (2) is converted to a percentage of irrigation water demand (WD), and Eq. (3) is introduced as vulnerability in this study:

$$\upsilon(\%) = \frac{1}{N-S}\sum_{i=1}^{N-S}\frac{D_i}{\mathrm{WD}} \times 100. \quad (3)$$

To determine long-term performance criteria realistically, stochastic demand for and stochastic supply of water should be taken into account (Dudley et al., 1972). However, in this study, as the main target is to assess the performance of the 3- and 2-year rotation systems, hereafter abbreviated as 3YR and 2YR systems, only the supply is considered stochastic. The demand is assumed as deterministic for two reasons: (i) the area is dominated by a single crop (sorghum) and (ii) the field is irrigated only once (before planting).

In addition to the annual volume of river flow, two parameters, namely intake ratio and irrigation efficiency, are considered for estimating the irrigation water supply. Intake ratio, as used here, refers to the fraction of river flow diverted to the GDSIS (volume of water diverted divided by total volume of river flow). The water allocation policies and practices of the GDSIS are such that 30 to 50 % of the river flow is diverted for spate irrigation (Anderson, 2011; IFAD, 2014). Moreover, the intake ratio is also a function of the diversion structure capacity (e.g. total number of gates) and operation (e.g. number of gates opened during irrigation). In this paper, irrigation efficiency refers to the ratio of the average amount of water stored in the root zone to the average amount diverted from the river. Annual silt removal from canals and the redesigning of the canals to adjust to local changes and regimes are some of the factors that affect the irrigation efficiency. The typical intake ratio and irrigation efficiency in the study area are 40 % (Anderson, 2011) and 50 % (Abdelgadir, 2007).

The water diverted from the river via the diversion structures undergoes conveyance and application losses. The net irrigation water supply (V_s) is therefore a fraction of the diverted amount and is expressed as follows:

$$V_s = \mathrm{IE} \times \mathrm{IR} \times V_r, \quad (4)$$

where V_r is the annual river flow (10^6 m^3), IR is the intake ratio (decimal) and IE is the irrigation efficiency (decimal).

In this study, sorghum, the dominant crop based on the current cropping pattern in the GDSIS (Anderson, 2011), is considered. According to Doorenbos and Kassam (1979), the net crop water requirements of sorghum with a total growing period of 110 to 130 days are between 450 and 650 mm. In spate irrigation in Sudan, an average of 500 mm net stored in the soil is used as the norm with a single watering per season (van Steenbergen et al., 2010). Hence, the net annual irrigation water demand for the 3YR (33 000 ha) and 2YR (50 000 ha) systems is about 165×10^6 and 250×10^6 m^3 respectively.

Based on the above-mentioned conditions, a Monte Carlo simulation was performed on the evaluation of reliability and vulnerability using generated river flow. The whole simulation process, which was written in R programming language, can be summarised as follows.

Step 1 Ensembles of 100-year river flow were generated using the two-parameter Weibull distribution ignoring autocorrelation.

Step 2 Each river flow value was converted to net irrigation water supply using Eq. (4).

Step 3 Each net irrigation water supply value was compared with the irrigation water demand to determine the number of satisfactory states (S) and deficits (D_i); then the reliability and vulnerability were computed using Eqs. (1) and (3) respectively.

Step 4 Steps 1 to 3 were repeated 5000 times to determine the average and distribution of the selected criteria.

3 Results and discussion

3.1 Reliability

Initially, the Monte Carlo simulation was performed for the 3YR and 2YR systems at typical values of intake ratio (40 %) and irrigation efficiency (50 %). The results, which reveal the distribution of irrigation water supply reliability of the 3YR and 2YR, are shown in Fig. 6. The mean reliability of the 3YR system is about 54 %. This result is consistent with the qualitative assessment on the performance of the GDSIS by Ngirazie et al. (2015), which showed that the irrigation water supply is reliable to 53 % of the farmers on average. When the annual target irrigated area is 50 000 ha (2YR), the mean reliability decreases to 18 %. The low value of the reliability for the 2YR reinforces the findings of other evaluation studies (e.g. Anderson, 2011; IFAD, 2012; IFAD, 2014) on the GDSIS rehabilitation, which state that the target area of 50 000 ha was never realistic. As a result, the target irrigated area was changed from 50 000 back to 33 000 ha in 2008 according to the midterm progress report (IFAD, 2012).

The coefficients of variation for the reliabilities of the 3YR and 2YR are 0.09 and 0.21 respectively. This implies that the reliability of the 2YR system is not only very low, but also more uncertain. Therefore, in terms of reliability, the 3YR is much better than the 2YR.

To test the sensitivity of the reliability to changes in the intake ratio (IR) and irrigation efficiency (IE), the Monte Carlo simulation was re-run for combinations of IR and IE. The resulting mean reliabilities are plotted in Fig. 7, which indicates that the reliability increases as the IR and IE increase. Although this trend is apparent without such analysis, the rate and amount of change is of practical interest for planners and decision makers. The mean reliability of the 3YR and 2YR systems for all IR and IE combinations varied from 0 to 91 % and from 0 to 77 % respectively.

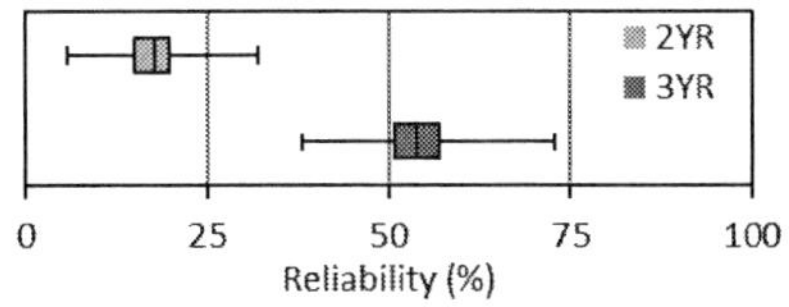

Figure 6. Box–whisker plots of the reliabilities of the 3YR (33 000 ha) and 2YR (50 000 ha) systems for typical values of intake ratio (IR) and irrigation efficiency (IE).

Relatively high reliability, greater than 60 %, can be attained for the 3YR even for the low IR of 30 % provided that the management of the system is good enough to achieve an IE of 70 %. This high value of reliability at a low IR may encourage planners and decision makers to invest more in improving the irrigation efficiency. However, such high reliability (> 60 %) is unattainable in the 2YR unless the IR is greater than 50 % for the same IE of 70 %. When the IR is very low, in this case 20 %, the reliability is less than 25 and 10 % for the 3YR and 2YR respectively for all IEs considered.

For the typical value of IE (50 %), the mean reliability of the 3YR and 2YR varied from 3 to 80 % and from 0 to 54 % respectively. When the emphasis is on maintaining higher reliability, say 75 % (the minimum standard value of irrigation system design in many countries), the IR has to be increased to more than 55 % in the 3YR system. This is unachievable in the 2YR even at the high IR of 60 % under the typical IE. When the IR is kept constant at its typical value (40 %), the mean reliability of the 3YR and 2YR varied from 10 to 77 % and from 0 to 48 % respectively. The 75 % reliability can only be achieved at the optimistically high IE of 70 % and only for the 3YR.

Referring to Fig. 7, the reliability values of 75 and 50 % for the 33 000 and 50 000 ha respectively can be achieved either by diverting 60 % of the river flow or managing the irrigation system to a level of 70 % IE. For both options, the increase in reliability would be accompanied by an increase in cost. However, the cost incurred for improving the IE can be compensated for by the value of the saved water. In fact, conserving water by improving the IE in irrigated agriculture can enhance equity among users. In contrast, the cost incurred by increasing the IR (through the expansion of irrigation structures) would be accompanied by reduced water availability to other sectors.

3.2 Vulnerability

As explained in Sect. 2.2, the Monte Carlo simulation of the vulnerability was conducted simultaneously and at the same IR and IE values as that of the reliability. The equivalent results are shown in Figs. 8 and 9. Figure 8 presents the distribution of the simulated vulnerabilities for the 3YR and 2YR systems for typical values of IR (40 %) and IE (50 %). As

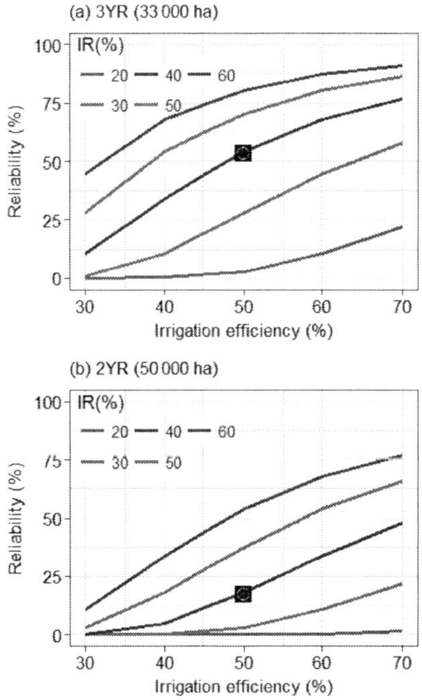

Figure 7. Mean reliability vs. irrigation efficiency (IE) at different levels of intake ratio (IR) of the 3- and 2-year rotation systems (the black square indicates typical values of IR and IE).

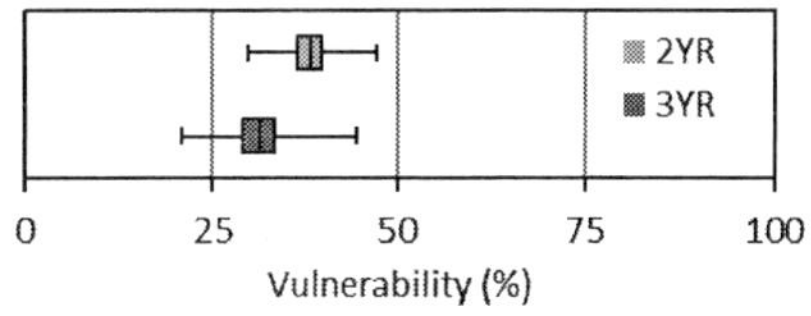

Figure 8. Box–whisker plots of the vulnerabilities of the 3YR (33 000 ha) and 2YR (50 000 ha) systems for typical values of intake ratio (IR) and irrigation efficiency (IE).

expected, the 2YR (irrigating 50 000 ha) is more vulnerable than the 3YR (irrigating 33 000 ha). The mean vulnerabilities of the 3YR and 2YR are 31 and 38 % respectively.

The mean vulnerabilities of the 3YR and 2YR for all IR and IE combinations, as shown in Fig. 9, varied from 25 to 67 % and from 28 to 77 % respectively. Figure 9 shows that the mean vulnerability changes marginally with increasing IE when IR is greater than 30 and 40 % for the 3YR and 2YR respectively. In both systems, the change in vulnerability is high for the 20 % IR. This implies that for the small amount of water diverted when the irrigation efficiency is kept very low, e.g. 30 %, the vulnerability is about 65 %. If efforts were made to improve the IE to e.g. 60 %, the vulnerability would be improved to 40 % (25 % improvement). In contrast, for

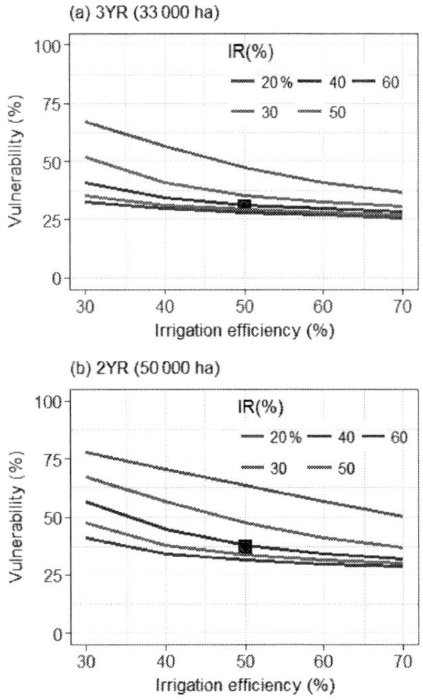

Figure 9. Mean vulnerability vs. irrigation efficiency (IE) at different levels of intake ratio (IR) of the 3- and 2-year rotation systems (the black square indicates typical values of IR and IE).

larger IR (e.g. 50 %), the corresponding change is only about 2 %. This implies that increasing the IR from 40 to 50 % or 60 % when the IE is greater than 40 % does not improve the vulnerability significantly.

An important observation from Figs. 7 and 9 is that the difference in vulnerabilities of the 3YR and 2YR is small compared to their respective reliability differences. For example, for an IR of 40 % and IE of 50 %, the reliability difference is 36 %, while the vulnerability difference is only 7 %. This indicates that in many years, the river flows, which are satisfactory in the 3YR, are marginally unsatisfactory in the 2YR. For example, if one chooses to irrigate 50 000 ha instead of 33 000 ha, the increment in the average deficit of water is only 7 %. This result, despite the high frequency of failure, might encourage the irrigation of 50 000 ha to accommodate many farmers. However, from an economic and sustainability point of view, it is necessary to carry out further investigation to translate the water-based vulnerability into a yield-based vulnerability. In other words, irrigating 50 000 ha instead of 33 000 ha could be justified if the increase in total yield from the increase in area by 50 % compensated for the reduction in yield caused by the relative average 7 % water deficit.

4 Conclusions

This study provides a simple methodology to assess the performance of the Gash Delta Spate Irrigation System by handling the interdependent relationships among the main variables of the system. The selected criteria are reliability and vulnerability. The criteria of the 3- and 2-year rotation systems were examined at typical values of the intake ratio (IR) and irrigation efficiency (IE), and the result showed that the 3-year rotation is preferable in the GDSIS. Moreover, the analysis in the feasible ranges of IR and IE revealed that there could be substantial improvement in reliability and vulnerability. This study showed how the rate and amount of change in the selected criteria are affected by individual and concurrent changes in the decision factors. The application of the methodology would provide planners and decision makers with guidance in the selection of a viable rotation system under different irrigation water supply and demand management strategies.

Data availability. Annual flow data of the Gash River measured at the Kassala Bridge gauging station were provided to the authors for this study on the GDSIS from the Gash Agricultural Scheme administrative office and the Gash River training unit, Kassala, Sudan. Currently, the data are not publicly accessible according to the data management policy of the Sudanese government.

Competing interests. The authors declare that they have no conflict of interest.

Special issue statement. This article is part of the special issue "Water security and the food–water–energy nexus: drivers, responses and feedbacks at local to global scales". It is a result of the IAHS Scientific Assembly 2017, Port Elizabeth, South Africa, 10–14 July 2017.

Acknowledgements. The authors would like to thank the Gash Agricultural Scheme administrative office and the Gash River training unit for sharing with us the necessary reports on Gash River flow data. This study was supported by the Grant-in-Aid for Scientific Research (B) from the Japan Society for the Promotion of Research (no. 16H04996, HT).

Edited by: Barry Croke
Reviewed by: three anonymous referees

References

Abdelgadir, K.: Evaluation of Field Irrigation of Gash Agricultural Scheme, MSc thesis, Water Management and Irrigation Institute, University of Gezira, Wad Medani, Sudan, 2007.

Anderson, I.: Technical Paper on Main Findings and Recommendations, The Easter Sudan Rehabilitation and Development Fund (ESRDF): GAS Phase II-Design Mission, ESRDF, Sudan, 2011.

Asefa, T., Clayton, J., Adams, A., and Anderson, D.: Performance evaluation of a water resources system under varying climatic conditions: Reliability, Resilience, Vulnerability and beyond, J. Hydrol., 508, 53–65, https://doi.org/10.1016/j.jhydrol.2013.10.043, 2014.

Cole, R.: Land tenure, agricultural labour, drought and food stress in the Gash, Gash Dai and Tokar agricultural areas: Measuring drough and drought impacts in Red Sea Province, Oxfam, Sudan, 1989.

Doorenbos, J. and Kassam, A.: Yield response to water, Rome, FAO, Irrigation and Drainage Paper 33, 1979.

Dudley, N. J., Musgrave, W. F., and Howell, D. T.: Irrigation planning 3. The best size of irrigation area for a Reservour, Water Resour. Res., 8, 7–17, https://doi.org/10.1029/WR008i001p00007, 1972.

Hashimoto, T., Stedinger, J. R., and Loucks, D. P.: Reliability, Resilience, and Vulnerability Criteria for Water Resource System Performance Evaluation, Water Resour. Res., 18, 14–20, https://doi.org/10.1029/WR018i001p00014, 1982.

IFAD: Gash Sustainable Livelihoods Regeneration Project (GSLRP): Target Group and Project Description, IFAD, Rome, 2004.

IFAD: Gash Sustainable Livelihoods Regeneration Project (GSLRP): Project completion report, Kassala, The republic of Sudan Ministry of Agriculture and Irrigatioin, Sudan, 2012.

IFAD: Republic of The Sudan: Gash Sustainable Livelihoods Regeneration Project, Project Performance Assessment, IFAD, 2014.

Jain, S. K.: Investigating the behaviour of statistical indices for performance assessment of a reservoir, J. Hydrol., 391, 90–96, https://doi.org/10.1016/j.jhydrol.2010.07.009, 2010.

Jain, S. K. and Bhunya, P. K.: Reliability, Resilience and Vulnerability of a Multipurpose Storage Reservoir, Hydrolog. Sci. J., 53, 434–447, https://doi.org/10.1623/hysj.53.2.434, 2008.

James, L.: Principles of farm irrigation systems design, John Wiley and Sons Limited, UK, 1988.

Kirkby, J.: Saving the Gash Delta, Sudan, Land Degrad. Dev., 12, 225–236, https://doi.org/10.1002/ldr.435, 2001.

Kjeldsen, T. R. and Rosbjerg, D.: Choice of reliability, resilience and vulnerability estimators for risk assessments of water resources systems, Hydrolog. Sci. J., 49, 755–767, https://doi.org/10.1623/hysj.49.5.755.55136, 2004.

Loucks, D. P. and van Beek, E.: Water Resources Systems Planning and Management: an introduction to methods, models and applications, UNESCO, Italy, 2005.

Ngirazie, L. A., Bushara, A. I., and Knox, J. W.: Assessing the performance of water user associations in the Gash Irrigation Project, Sudan, Water Int., 1–12, https://doi.org/10.1080/02508060.2015.1072677, 2015.

Swan, C.: The Recorded Behaviour of the River Gash in Sudan, Ministry of Irrigation and Hydro-electric Power, Khartoum, Sudan, 1959.

van Steenbergen, F.: Understanding the sociology of spate irrigation: cases from Balochistan, J. Arid Environ., 35, 349–365, https://doi.org/10.1006/jare.1996.0171, 1997.

van Steenbergen, F., Haile, A. M., Lawrence, P., Salman, M., and Faurès, J.: Guidelines on Spate Irrigation, Rome, FAO, Irrigation and Drainage Paper 65, 2010.

Zimmerman, J.: Irrigation, John Wiley & Sons, Inc, New York, 1966.

Proc. IAHS, 376, 77–82, 2018
https://doi.org/10.5194/piahs-376-77-2018

Open Access
PIAHS

Modelling the impact of mulching the soil with plant remains on water regime formation, crop yield and energy costs in agricultural ecosystems

Yeugeniy M. Gusev, Larisa Y. Dzhogan, and Olga N. Nasonova

Laboratory of Soil Water Physics, Institute of Water Problems, Russian Academy of Sciences, 119333 Moscow, Russian Federation

Correspondence: Yeugeniy M. Gusev (sowaso@yandex.ru)

Received: 31 May 2017 – Accepted: 20 August 2017 – Published: 1 February 2018

Abstract. The model MULCH, developed by authors previously for simulating the formation of water regime in an agricultural field covered by straw mulch layer, has been used for the comparative evaluation of the efficiency of four agricultural cultivation technologies, which are usually used for wheat production in different regions of Russia and Ukraine. It simulates the dynamics of water budget components in a soil rooting zone at daily time step from the beginning of spring snowmelt to the beginning of the period with stable negative air temperatures. The model was designed for estimation of mulching efficiency in terms of increase in plant water supply and crop yield under climatic and soil conditions of the steppe and forest-steppe zones. It is used for studying the mulching effect on some characteristics of water regime and yield of winter wheat growing at specific sites located in semi-arid and arid regions of the steppe and forest-steppe zones of the eastern and southern parts of the East-European (Russian) plain. In addition, a previously developed technique for estimating the energetic efficiency of various agricultural technologies with accounting for their impact on changes in soil energy is applied for the comparative evaluation of the efficiency of four agricultural cultivation technologies, which are usually used for wheat production in different regions of the steppe and forest-steppe zones of the European Russia: (1) moldboard tillage of soil without irrigation, (2) moldboard tillage of soil with irrigation, (3) subsurface cultivation, and (4) subsurface cultivation with mulching the soil with plant remains.

1 Introduction

The present work considers the possibility of rational use of natural resources (water, soil, and energy), which are interconnected, in particular, through food-water-energy nexus. One of non-traditional agricultural technologies, which are developing in many countries and oriented on sustainable management, on conservation of soil, energy and water resources, as well as on protection of environment, is soil mulching, i.e. covering the soil surface with plant remains after previous harvesting. However, application of this agricultural technology is mainly based on an empirical approach without estimating its efficiency, which would allow one to systematize the obtained results and predict the implications of mulching (Scopel et al., 2004; Balwinder-Singha et al., 2016). This raises the problem of developing the universal technique for calculating an increase in water supply and yield of agroecosystems due to mulching as well as predicting the efficiency of mulching under different natural conditions. This work is intended to contribute to the solution of the stated problem. The aim of the present study is development of a technique for calculating the efficiency of soil mulching and application of the developed technique for agricultural fields located in the eastern and southern parts of the East-European (Russian) plain.

One more objective of the study is application of a previously developed technique for evaluating the energetic efficiency of different agricultural management practices with accounting for their effect on soil energy. Here, the technique is used for comparative evaluation of the efficiency of four agricultural technologies: (1) moldboard tillage without irrigation, (2) moldboard tillage with irrigation, (3) subsurface

Published by Copernicus Publications on behalf of the International Association of Hydrological Sciences.

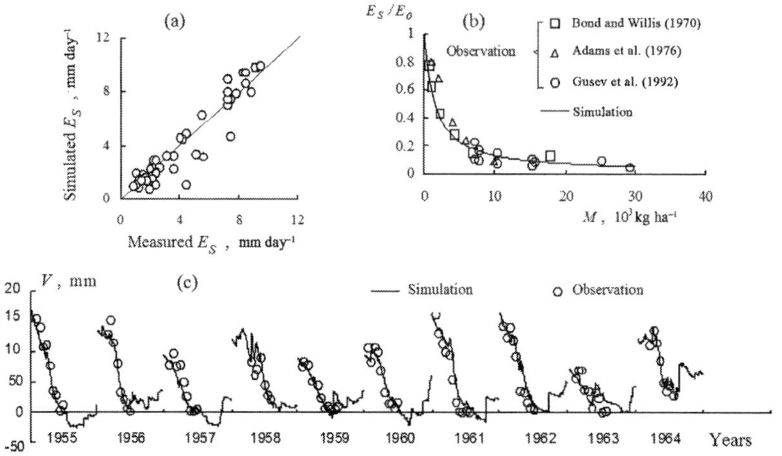

Figure 1. Model validation: **(a)** comparison of simulated daily bare soil evaporation and measured one at the Kursk Biospheric station (solid line is the 1 : 1 line); **(b)** cumulative (for some rainless intervals) evaporation from bare soil covered with straw mulch vs. surface density of straw M (E_0 is the potential bare soil evaporation without mulch cover); **(c)** dynamics of simulated and measured productive water storage V in 1 m soil layer during the warm seasons (from April to October) in wheat fields at the Ershov agrometeorological station.

cultivation, and (4) subsurface cultivation with mulching the soil with plant remains, which are usually used for wheat production at different agrometeorological stations located in the steppe and forest-steppe zones of the European Russia.

2 Brief description of the MULCH model

Description of the MULCH model, which simulates the formation of water regime in an agricultural field covered with a straw mulch, is presented in (Gusev and Dzhogan, 2000; Gusev and Busarova, 2001; Gusev, 2012). The model calculates the dynamics of water budget components in a soil rooting layer (taken to be equal to 1 m) at daily time step from the beginning of spring snowmelt to the beginning of the period with stable negative air temperatures. It treats the following processes: interception of precipitation by the mulch layer and subsequent evaporation, transpiration, evaporation from the soil under the mulch layer, water exchange with the underlying soil horizons, and changes in water storage of the sol rooting zone.

The model is intended for estimation of mulching efficiency in terms of increase in plant water supply and crop yield under different climatic and soil conditions.

Interception of precipitation by the mulch layer is evaluated using the so-called "bucket" concept. In so doing, the mulch serves as an intercepting reservoir of a constant capacity. Calculation of the water budget components within a soil–mulch–vegetation–atmosphere system is based on the solution of the following water budget equation:

$$\frac{\mathrm{d}V}{\mathrm{d}t} = P - \mathrm{ET} - Q - Y, \qquad (1)$$

$$\mathrm{ET} = E_\mathrm{T} + E_\mathrm{S} + E_\mathrm{m}, \qquad (2)$$

where t is the time, V is the productive soil water storage in the soil rooting layer (equaled to the difference between the total soil water storage and plant wilting point), P is the rainfall, ET is the evapotranspiration, E_T is the transpiration, E_S is the bare soil evaporation, E_m is the evaporation from the mulch layer, Q is the water exchange at the lower boundary of the root zone, and Y is the surface runoff. Technique for the calculation of the components E_T, E_S and E_m is described in a number of publications (e.g. Budagovskii, 1989; Gusev, 1998; Gusev and Dzhogan, 2000).

Along with development of the calculational algorithm, preparation of input data represents an important stage of elaborating the model. Inputs for the MULCH model include (1) forcing data (precipitation, wind speed, air temperature and humidity, net radiation), (2) initial spring soil water storage, (3) soil hydrophysical parameters, (4) mulch parameters, and (5) type of vegetation.

3 Validation of the MULCH model

The validation of the MULCH model was performed using different experimental data. First of all, the ability of the model to reproduce bare soil evaporation E_S without mulch layer was investigated, because mulching substantially affects E_S and it is very important to be sure that E_S is simulated adequately. Figure 1a shows a comparison of simulated daily E_S with corresponding measured values. The measurements were performed by means of specific evaporometers at the Kursk Biospheric Station (Institute of Geography, Russian Academy of Sciences) located in the forest-steppe zone

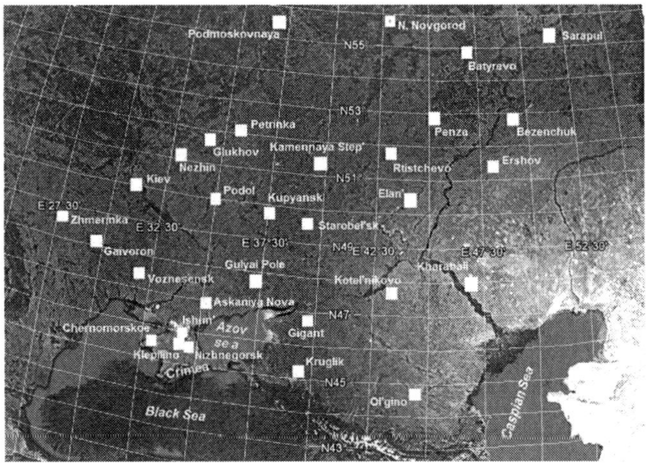

Figure 2. The study region (the southern part of the Russian Plain). The white squares are agrometeorological stations.

and characterized by a typical chernozem soil. The comparison of simulated and measured data demonstrates the possibility of application of the developed model for simulating bare soil evaporation.

At the next stage, the ability of the MULCH model to simulate evaporation from the soil covered with plant mulch under unlimited water supply (i.e. in the absence of dry layer at the soil surface) was investigated. Some results are given in Fig. 1b. The measurements were taken from literature (Bond and Willis, 1970; Adams et al., 1976) and obtained in our field experiments (Gusev et al., 1992).

Besides, the model was tested for agricultural ecosystems without mulch and while studying the mulching effect on some characteristics of the water regime and yield of winter wheat in several sites located in semi-arid and arid regions of the steppe and forest-steppe zones situated in the eastern and southern parts of the East-European (Russian) plain. An example of dynamics of soil water storage in 1 m soil layer without mulching as simulated by the MULCH model and measured at the agrometeorological station Yershov, which is located in the arid region of the steppe zone, is shown in Fig. 1c.

Analysis of the results of model validation allowed us to conclude that the proposed model can be used for the solution of the problem of estimating the efficiency of soil mulching in terms of increase in water supply of agricultural ecosystems.

4 The study area and data

In the present work, the territory under consideration includes the south of the European part of Russia, the Crimea and the south-eastern part of Ukraine (Fig. 2). This region situated in the steppe and forest-steppe zones is characterized by arid and semi-arid climate. Therefore, the highest efficiency of mulching can be expected there. In such a region, as a rule, the summer surface runoff Y and upward components of water exchange between the soil root zone and the underlying layers Q can be neglected that allowed us to simplify the model and use the following equation instead of Eq. (1):

$$\frac{\mathrm{d}V}{\mathrm{d}t} = P - \mathrm{ET}. \tag{3}$$

The records of standard observations from agrometeorological stations (Fig. 2) were used as meteorological forcing data and initial values of soil water storage. The soil hydrophysical parameters were also derived from the results of field studies performed at the stations. The dynamics of a leaf area index of wheat was calculated with the help of the model for its formation described in Gusev and Busarova (1998). This model suggests that evolution of the leaf area index of a specific crop during the vegetation period is determined by some preset ("genetic") program, dynamics of air temperature, and water regime of a soil root zone.

The parameters of mulch were determined during the field experiments (Gusev et al., 1992). It was found that the convective diffusivity of water vapor in a mulch layer is a function of the wind speed; the water-holding capacity of the mulch consisting of fresh plant remains can be taken as $1.8\,\mathrm{kg\,kg^{-1}}$. The surface density of the mulch layer, which also determines its depth under the given bulk density of straw (of about $15–20\,\mathrm{kg\,m^{-3}}$), is a varying parameter, which depends on the facilities at each farm. It can reach $0.1–1.0\,\mathrm{kg\,m^{-2}}$, which results in ranging of the depth of a mulch layer between 0.5 and 5.0 cm at the given bulk density of straw.

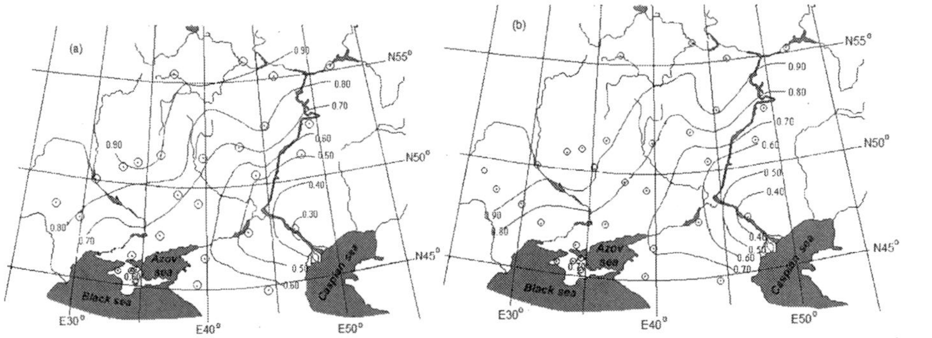

Figure 3. Distribution of many-year averages of the index of plant water supply η_w **(a)** without mulching the agricultural fields and **(b)** in the case of soil mulching.

5 Indexes of plant water supply in agroecosystems

The following indexes can be used for comparison of changes in water regime of agricultural system due to influence of different agrotechnologies (Budagovskii, 1985; Gusev and Novak, 2007):

$$\eta_w = E_T / E_{TP}, \quad (4)$$

$$\eta_e = E_T / \mathrm{ET}, \quad (5)$$

where E_{TP} is the potential transpiration, ET is the evapotranspiration. The former index (Eq. 4) characterises plant water supply, while the latter one (Eq. 5) can be treated as an efficiency of utilization of soil water resources by vegetation. Calculation of these indexes is based on the modelling of water exchange processes within a soil–mulch–vegetation–atmosphere system (Gusev and Dzhogan, 2000). To determine the efficiency of soil mulching these indexes and winter wheat yield were calculated both in the presence and absence of mulch layer.

6 Estimates of the efficiency of mulching

A comparison of the indexes of crop water supply and the difference in crop yield in the presence and absence of mulch layer made it impossible to evaluate the efficiency of mulching. Model simulations were performed for 20–25 years for winter wheat. Analysis of the results showed that mulching results in noticeable changes in water regime of agricultural fields and crops yield until the thickness of mulch layer reaches 4–5 cm. Further increase in the mulch thickness does not practically influence these characteristics. Thus, there is some optimal thickness of the mulch layer ($\sim$5 cm), which allows one to improve plant water supply and increase crop yield. This is why, in further analysis, the thickness of mulch layers is taken to be equal to 5 cm.

The performed calculations made it possible to illustrate the spatial distribution of plant water supply η_w in the case of application of modern technology of cultivation of winter wheat both in the absence of mulching (Fig. 3a) and with mulching (Fig. 3b). (The limited volume of the article does not allow us to consider the index η_e.) As can be seen in Fig. 3a, the index η_w has a pronounced tendency to decrease towards the southeast and varies from 0.90 to 0.30 over the territory when a traditional agrotechnology is used. In the most important for wheat production forest-steppe and steppe regions, it equals to 0.60–0.70 and 0.70–0.80 in the Russian and in the Ukrainian part of the region, respectively. Application of the mulch layer leads to increase in η_w by 10–15 %. Such an increase is very important, in particular, for the steppe part of Crimea, because in 2014 the Ukrainian government blocked the North-Crimean Channel, which largely provided irrigation for agroecosystems.

Figure 4 shows a spatial distribution of mean increase in winter wheat yield due to mulching. As can be seen in Fig. 4, in the forest-steppe and steppe zones the increase mainly equals to 3–5 centner ha^{-1}, changing from 6–7 centner ha^{-1} in the more humid north-western and northern areas to 2 centner ha^{-1} in the southern arid regions. Experimental estimates of the impact of mulching on the yield of cereals are very limited. More reliable data are given in Lomakin (1988). According to Lomakin (1988), the increase in the yield of grain crops equals, on average, to 4.5 centner ha^{-1} in Petrinka (see its location in Fig. 2) that is in a good agreement with our results.

7 Energetic criteria of the efficiency of agricultural technologies

In addition, the energetic efficiency of various agricultural technologies, such as: (1) mouldboard tillage of soil without irrigation, (2) mouldboard tillage of soil with irrigation, (3) subsurface (reduced) cultivation, and (4) subsurface cultivation with mulching the soil with plant remains, was estimated using a technique developed in Gusev and Busarova (2001). These agricultural cultivation technologies

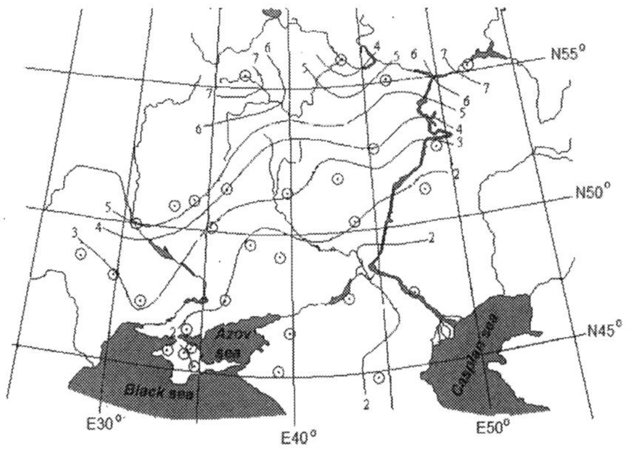

Figure 4. The increase in many-year averages of winter wheat productivity (in centner ha^{-1}) due to mulching the soil with plant remains.

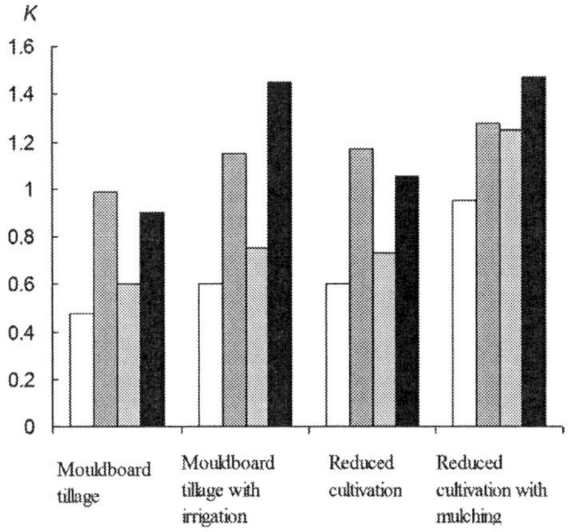

Figure 5. Energetic efficiency of various tillage modes with taking into account the change in soil energy for wheat production at the following agrometeorological stations: Ershov (empty boxes), Bezenchuk (dark grey boxes), Kamennaya Step' (light grey boxes), and Petrinka (black boxes).

are usually used for wheat production in different regions of the steppe and forest-steppe zones of the European part of Russia. In so doing, impact of these technologies on soil energy was also taken into account, because agricultural technologies result in changes in the content of organic matter (humus) in the soil. The content of humus largely determining soil fertility and influencing a primary organic production is directly connected with soil energy. Thus, to estimate the efficiency of an agricultural technology, it is necessary to add a component describing changes in soil energy to an energetic criterion used for comparison of agrotechnologies:

$$K = \frac{\mathrm{En}_{\mathrm{out}} + \Delta \mathrm{En}_{\mathrm{s}}}{\mathrm{En}_{\mathrm{in}}}, \tag{6}$$

where $\mathrm{En}_{\mathrm{in}}$ is incoming energy from various anthropogenic sources, $\mathrm{En}_{\mathrm{out}}$ is the output energy of yield in the form of chemical energy of food and raw organic matter, $\Delta \mathrm{En}_{\mathrm{s}}$ is the change in soil energy during a corresponding time interval (for example, a year). The total anthropogenic energy $\mathrm{En}_{\mathrm{in}}$ consists of energetic expenses for crop production and ingathering, for manufacture and operation of engineering and equipment, their amortization, for capital investments, irrigation, production of fuel, mineral fertilizers, etc.

The technique for estimating all energetic components in Eq. (6) was developed in Gusev and Busarova (2001). Here, we give an example of estimation of K for comparison of the abovementioned agricultural tillage modes, which were used for wheat production at four agrometeorological stations located in the European part of Russia: Petrinka, Kamennaya Step', Bezenchuk, and Ershov (Fig. 5).

The obtained estimates of the energetic efficiencies of the tillage modes allowed us to conclude that mulching the soil with plant remains in combination with reduced (subsurface) cultivation is the most effective agricultural mode for the regions under study. This technology can be considered as the most promising for the development of agriculture at least in these regions of Russia.

In Kolganov et al. (2016) it is believed that about 10 % of arable land in water-deficit and arid regions of Russia need to be irrigated to provide higher and more stable yields in agriculture (these regions are considered in the present paper). Currently, approximately 4.2 % of arable land in Russia is irrigated.

We have considered two techniques for increasing yields of wheat for the Russian part of the region (including the Crimea), considered in the present paper: (1) increasing the share of irrigated land in the region to by 10 %, (2) obtaining the same annual gross yield on the same arable land, using the reduced cultivation with mulching (instead of increasing irrigated areas). It turned out that the same (for the two considered techniques) annual gross yields of wheat (6.5×10^8 GJ yr^{-1}, this is equivalent to the yield of grain of ~ 43 million t yr^{-1}) are obtained if in the second case the reduced cultivation with mulching is used for 30 % of arable land. Herewith the annual energetic costs in the second case are less than in the first case by 10^8 GJ yr^{-1}. If this difference in energy costs is translated into the cost of the volume of oil (with the same energy content), then, at an oil price of nearly USD 50 per barrel, the second technique is cheaper than the first one approximately by USD 1 billion yr^{-1}.

8 Conclusion

1. The MULCH model describing formation of the water regime of agroecosystems allows one to estimate the impact of soil mulching on plant water supply and crop yield. The model was applied for estimating the effi-

ciency of mulching the winter wheat fields located in the eastern and southern parts of the East-European (Russian) plain.

2. It is shown that mulching with plant remains results in noticeable changes in water supply and yield of winter wheat only if the thickness of mulch layer is within 4–5 cm. Further increase in the thickness does not practically influence these characteristics.

3. Mulching with plant remains leads to increase in the ratio of actual transpiration to potential one, on average, by 10–15 % over the study area. At the same time the increase in the yield of winter wheat is 3–5 centner ha^{-1} for most of the territory varying from 6–7 centner ha^{-1} in more humid north-western and northern regions to 2 centner ha^{-1} in arid southern areas.

4. The obtained estimations of energetic efficiency for the four tillage modes usually used for wheat production allow us to conclude that soil mulching with plant remains in combination with subsurface cultivation is the most efficient agricultural mode for the regions of the steppe and forest-steppe zones of the East-European (Russian) plain. This technology can be considered as the most promising for development of agriculture in these regions, because it does not require an increase in traditional resources (for example, irrigation) used there.

Data availability. Observations from agrometeorological stations can be taken from the Russian Research Institute of Hydrometeorological Information – World Data Center (RIHMI-WDC) (http://meteo.ru).

Competing interests. The authors declare that they have no conflict of interest.

Special issue statement. This article is part of the special issue "Water security and the food–water–energy nexus: drivers, responses and feedbacks at local to global scales". It is a result of the IAHS Scientific Assembly 2017, Port Elizabeth, South Africa, 10–14 July 2017.

Acknowledgements. The study related to the eastern and southern parts of the East-European (Russian) plain was supported by the Russian Science Foundation (grant no. 16-17-10039), the study related to the Crimea was supported by the Russian Academy of Sciences.

Edited by: Barry Croke
Reviewed by: two anonymous referees

References

Adams, J. E., Arkin, G. F., and Ritchie, J. T.: Influence of row spacing and straw mulch on first stage drying, Soil Sci. Soc. Am. J., 40, 436–442, 1976.

Balwinder-Singha, Humphreysb, E., Gaydond, D. S., and Eberbachf P. L.: Evaluation of the effects of mulch on optimum sowing date and irrigation management of zero till wheat in central Punjab, India using APSIM, Field Crops Res., 197, 83–96, 2016.

Bond, J. J. and Willis, W. O.: Soil water evaporation: first stage drying as influenced by surface residue and evaporation potential, Soil Sci. Soc. Am. Proc., 34, 924–928, 1970.

Budagovskii, A. I.: Soil water reserves and the water supply of plant cover, Vodn. Res., 4, 3–13, 1985.

Budagovskii, A. I.: Main principles of calculation of irrigation rates and irrigation regimes, Vodn. Res., 1, 38–48, 1989.

Gusev, E.: Soil water resources and ecology of terrestrial vegetation cover. Conceptions, experiment, calculations, Palmarium Academic Publishing, Saarbrücken, 2012.

Gusev, E. M.: Evaporation from soil under drying, Euras. Soil Sci., 8, 836–840, 1998.

Gusev, E. M. and Busarova, O. E.: Modeling the dynamics of the leaf area index for grain crops, Russ. Meteorol. Hydrol., 1, 70–75, 1998.

Gusev, E. M. and Busarova, O. E.: Evaluating the energy efficiency of agricultural technologies, Euras. Soil Sci., 34, 741–751, 2001.

Gusev, E. M. and Dzhogan, L. Y.: Evaluation of the effect of soil mulching with plant residues on the water regime of agroecosystems, Euras. Soil Sci., 11, 1238–1247, 2000.

Gusev, E. M., Busarova, O. E., Shurkhno, A. A., and Yasinskii, S. V.: The effect of straw mulch on the thermal regime of soil after snow melting, Pochvovedenie, 5, 49–59, 1992.

Gusev, Y. and Novák, V.: Soil water – main water resources for terrestrial ecosystems of the biosphere, J. Hydrol. Hydromech., 55, 3–15, 2007.

Kolganov, A. V., Suhoi, N. V., Shkura, V. N., and Shchedrin, V. N.: The development of reclamation of agricultural land in Russia, RosNIIPM, Novocherkassk, 2016.

Lomakin, M. M.: Mulch Soil Tillage on Slope, Agropromizdat, Moscow, 1988.

Scopel, E., Da Silva, F., Corbeels, M., Affholder, F., and Maraux, F.: Modelling crop residue mulching effects on water use and production of maize under semi-arid and humid tropical conditions, Agronomie, 24, 383–395, 2004.

Proc. IAHS, 376, 83–86, 2018
https://doi.org/10.5194/piahs-376-83-2018

Open Access PIAHS

Socio-Hydrological Approach to the Evaluation of Global Fertilizer Substitution by Sustainable Struvite Precipitants from Wastewater

Dirk-Jan Daniel Kok[1], Saket Pande[1], Angela Renata Cordeiro Ortigara[2], Hubert Savenije[1], and Stefan Uhlenbrook[1,2]

[1]Department of Water Management, Delft University of Technology, Delft, the Netherlands
[2]World Water Assessment Programme, UNESCO, Perugia, Italy

Correspondence: Dirk-Jan Daniel Kok (d.d.kok@student.tudelft.nl)

Received: 25 July 2017 – Accepted: 3 October 2017 – Published: 1 February 2018

Abstract. Despite Africa controlling the vast majority of the global phosphate it also faces the greatest food shortages – partially due to a lack of access to the fertilizer market. A more accessible source of phosphorus comes from wastewater flows, which is currently lost through the discharge to open surface waters. Analysing the potential phosphorus production of urban and livestock wastewater in meeting partial agricultural demand for phosphorus can improve food security, reduce consumption of unrenewable phosphorus, reduce pollution, and aid the transitioning to a circular economy. In this study, a global overview is provided where a selection of P-production and P-consumption sites have been determined using global spatial data. Distances, investment costs and associated carbon footprints are then considered in modelling a simple, alternative trade network of struvite phosphorus flows. The network reveals potential for increasing the phosphorus security through phosphorus recycling in particularly the South Africa, Lake Victoria and Nigeria regions. Given Africa's rapid urbanization, phosphorus recovery from wastewater will prove an important step in creating sustainable communities, protecting the environment while improving food security, and so contributing to the United Nations 2030 Agenda for Sustainable Development.

1 Introduction

Phosphorus (P) is an element necessary for the development of crops as it forms a key, structural component of DNA and RNA. It is applied in the form of single or triple superphosphate, or mono-Ammonium or di-Ammonium Phosphate (DAP) fertilizers which are both easy to transport, to distribute over fields and are readily absorbed by plants. The most essential source for the production of phosphorus fertilizer is phosphate ore. Some authors predict a peak production of phosphate ore could occur as early as 2030 (Cordell et al., 2009), or that extractable mineral P resources will become scarce or exhausted within the next 50 to 100 years (Steen, 1998; Smil, 2000; van Vuuren et al., 2010). This prospect threatens the food security situation of Sub-Saharan Africa (SSA), as nearly 75 % of SSA's agricultural soils are nutrient deficient and so already contributing significantly to the crop yield gaps (IFDC, 2006). The immediate results of phosphate rock depletion will be a further reduction of the accessibility to fertilizers by small-holder and sustenance farmers that comprise areas already coping with food shortages.

1.1 Potential to Recover

Despite the issues surrounding phosphorus demands and yield gaps, there currently exists no financially attractive recovery technology for the enormous phosphorus recovery potential from livestock (Schoumans et al., 2015), while urban recovery is often also limited due to financial constraints. The competitive position of the relatively expensive, recovered phosphorus is improving, however, as over the past 15 years the phosphorus price of di-Ammonium Phosphate (DAP; Index Mundi, 2017) has increased from 665 [USD t^{-1}] to 1552 [USD t^{-1}]. In that same period, the

Published by Copernicus Publications on behalf of the International Association of Hydrological Sciences.

price has been as high as 5217 [USD t^{-1}] during the economic crisis around 2008 and as low as 656 [USD t^{-1}] in 2002 (Index Mundi, 2017). Given these price trends, and that the will, the technology and the knowledge are there to facilitate the largescale implementation of phosphorus recovery, it is important that investigation is done as to what such a development may come to look like. This study aims to identify and connect those areas where there exists a high potential for the recovery of a sustainable phosphorus products from urban and livestock wastewater, to those areas of high agricultural demand.

2 Materials and Methods

Using population density maps (CIESIN, 2016; Robinson et al., 2014) and globally generalized phosphorus excretion rates (Gilmour et al., 2008; Barker et al., 2001; CBS, 2014), a crude mapping of phosphorus production sites is carried out at global scale. Apart from these sustainable sources, a total of 5.2 [mt a^{-1}] P-production from major African mines is included also (USGS, 2007). The demand for phosphorus is assessed through crop harvested area maps for 6 major crops that account for approximately 50–56 % of the global phosphorus fertilizer consumption (Heffer and Prud'homme, 2008). The P demands for maize, wheat, rice, sorghum, soy bean, and potato are approximated for the optimal, but water-constrained, yield as determined through the transpiration deficit method (Steduto et al., 2012). The associated phosphorus requirement for this yield is determined through a linear regression between yield and P-fertilizer application. These demands are then proportionally scaled up to match the global agricultural demand of all crops of 16 [Mt a^{-1}]. The remaining production and demand sites are connected to each other through a network. The creation process accounts for generalized struvite precipitation costs by Eq. (1),

$$f^i_{\mathrm{min}} = \frac{P_{\mathrm{m}} \times R_{\mathrm{mp}} + B^s}{S_{\mathrm{PT}}} - S_{\mathrm{s}} \tag{1}$$

where f^i_{min} is the minimum price for phosphorus for node i [USD t^{-1}]; P_{m} is the price of magnesium [USD t^{-1}]; R_{mp} is the ratio of magnesium required per ton of phosphorus; B^i fixed operational cost [USD]; S_{PT} is the total phosphorus production potential [t]; and S_{s} is the minimum scaling cost savings from struvite precipitation, 620 [USD t^{-1}] (Shu et al., 2006). Similarly, maximum prices that demand nodes are able to pay are determined roughly through Eq. (2).

$$f^n_{\mathrm{max}} = \frac{Y^n_{\mathrm{opt}} \times C^n_a}{P^n_{\mathrm{opt}}} \times R^n \tag{2}$$

where f^n_{max} is the max price for phosphorus [USD t^{-1}]; Y^n_{opt} is the optimal yield [t ha^{-1}]; C^n_a is the crop price in year a [USD t^{-1}]; P_{opt} is the optimum fertilizer dosage rate (equal to P-demand for optimal yield) [t ha^{-1}]; and R^n is the ratio of fertilizer cost to total costs [–], for crop n. The transportation costs are determined with *as-the-crow-flies* distances with a land transport cost equation as a basis. In this are considered: the price of diesel, a labour wage of 17 [USD h^{-1}], an average velocity of 80 [km h^{-1}], a load capacity of 60 tonnes (2 × 30), and a fuel efficiency of 0.53 [L km^{-1}] at capacity. The model is run for a future P-supply scenario of mine production supplemented by sustainable sources that are introduced individually when market prices make their recovery feasible.

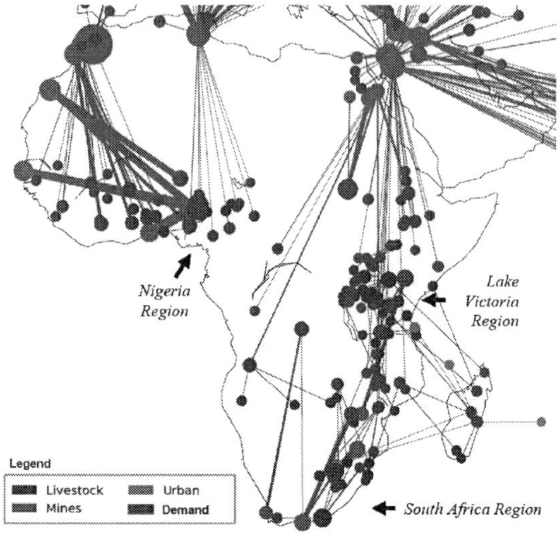

Figure 1. Dominant network pattern of potential trade in recovered P (on top of rock based P).

3 Results

While the network analysis is conducted globally for the year 2005, only outcomes relevant for continental Africa are discussed here. The network reveals potential for increasing the phosphorus security of particularly Rwanda, Burundi, Kenya, Tanzania, Uganda, Malawi, South Africa, and Nigeria, through phosphorus recycling (Fig. 1).

Approximate phosphorus production potentials of 130 [kt a^{-1}] and 1300 [kt a^{-1}] from concentrated urban and livestock areas respectively are determined for continental Africa. The phosphorus demand for optimal, water restricted yield in Africa for the six major crops is approximately 1 [mt a^{-1}], which equals 12 % of the *global* crop demand. Continental wide struvite precipitation from major *urban population centres* can theoretically then satisfy a maximum of 13 % of the agricultural demand for these crops. The Lake Victoria and South African regions show a higher than average density of fluxes in sustainable trade. The first areas to offer competitive sustainable phosphorus lie in these regions, entering the market around prices of 600 [USD t^{-1}] P. The African market in recoverables, i.e. recycled phosphorus, expands further at 800 [USD t^{-1}], and continues to grow in smaller amounts at higher prices.

4 Discussion and Conclusion

The model offers a simple framework for network assessment of optimal global phosphorus trade and maximum phosphorus potentials. Provided it is at global scale and that we have only discussed outcome relevant for Africa, however, the study is also superimposed with many generalisations and assumptions that introduce many side notes to its validation. Inaccuracies in assessing the urban and livestock production potentials due to generalization of throughput figures, the inconsideration for trans-Atlantic trade, the *as-the-crow-flies* distance method, assumptions of free trade, as well as the cost equation assuming land transport costs per kilometre – disregarding cheaper costs for shipping – all possibly contribute to an over approximation of the market prices.

Sustainable products become more competitive with higher market prices for phosphorus. Higher market prices are created as the price of fuel rises, mine exploitation costs increase and total supply decreases. Sustainable sources already have a potential to be competitive in South Africa, the Victoria Lake region and Nigeria given the close proximity of production and demand nodes of significant supply and demands, and their relative distance from the world's largest phosphate mines. The introduction of recovered products to the market will result in: (1) an increase in access to fertilisers within Africa and elsewhere, (2) lower phosphorus cost, (3) cause for a significant reduction in transport associated emissions, (4) stimulate the treatment of wastewater, and (5) contribute to improving food security.

Data availability. This study's produced datasets are no longer available with the updating of the model. The complete and updated datasets are available from the corresponding author on request.

Competing interests. The authors declare that they have no conflict of interest.

Special issue statement. This article is part of the special issue "Water security and the food–water–energy nexus: drivers, responses and feedbacks at local to global scales". It is a result of the IAHS Scientific Assembly 2017, Port Elizabeth, South Africa, 10–14 July 2017.

Edited by: Graham Jewitt
Reviewed by: Rebecca Sindall

References

Barker, J. C., Hodges, S. C., and Walls, F. R.: Livestock Manure Production Rates and Nutrient Content, 1–4, 2001.

CBS (Centraal Bureau voor de Statistiek): Dierlijke mest en mineralen 2013, The Hague: Centraal Bureau voor de Statistiek, available at: https://www.cbs.nl/-/media/imported/documents/2014/49/2014-dierlijke-mest-en-mineralen-2013-pub.pdf (last access: 11 October 2016), 2014.

CIESIN (Center for International Earth Science Information Network): Columbia University, Gridded Population of the World, Version 4 (GPWv4): Population Density, https://doi.org/10.7927/H4NP22DQ, 2016.

Cordell, D., Drangert, J.-O., and White, S.: The story of phosphorus: Global food security and food for thought, Glob. Environ. Change, 19, 292–305, https://doi.org/10.1016/j.gloenvcha.2008.10.009, 2009.

Gilmour, D., Blackwood, D., Comber, S., and Thornell, A.: Identifying human waste contribution of phosphorus loads to domestic wastewater, Proceedings at the 11th International Conference on Urban Drainage, Held in Edinburgh, Scoltand, UK, 1–10, https://web.sbe.hw.ac.uk/staffprofiles/bdgsa/11th_International_Conference_on_Urban_Drainage_CD/ICUD08/pdfs/740.pdf (last access: 27 November 2017), 2008.

Heffer, P. and Prud'homme, M.: Medium-Term Outlook for Global Fertilizer Demand, Supply and Trade 2008–2012, 76th IFA Annual Conference Vienna (Austria), 19–21 May 2008.

IFDC (International Center for Soil Fertility and Agricultural Development): gricultural Production and Soil Nutrient Mining in Africa: Implications for Resource Conservation and Policy Development, IFDC: Muscle Shoals, Alabama 35662, USA, 2006.

Index Mundi Commodity Price Indices: DAP Fertilizer, available at: http://www.indexmundi.com/commodities/, last access: 15 July 2017.

Robinson, T. P., William Wint, G. R., Conchedda, G., van Boeckel, T. P., Ercoli, V., Palamara, E., and Gilbert, M.: Mapping the global distribution of livestock, PLoS ONE, 9, e96084, https://doi.org/10.1371/journal.pone.0096084, 2014.

Schoumans, O. F., Bouraoui, F., Kabbe, C., Oenema, O., and van Dijk, K. C.: Phosphorus management in Europe in a changing world, Ambio, 44, 180–192, https://doi.org/10.1007/s13280-014-0613-9, 2015.

Shu, L., Schneider, P., Jegatheesan, V., and Johnson, J.: An economic evaluation of phosphorus recovery as struvite from digester supernatant, Bioresource Technol., 97, 2211–2216. https://doi.org/10.1016/j.biortech.2005.11.005, 2006.

Smil, V.: Phosphorus in the environment: Natural flows and human interferences, Annu. Rev. Energ. Env., 25, 53–88, https://doi.org/10.1146/annurev.energy.25.1.53, 2000.

Steen, I.: Management of a non-renewable resource, Phosphorus and Potassium, 217, 25–31, 1998.

Steduto, P., Hsiao C., T., Fereres, E., and Raes, D. FAO Irrigation and Drainage Paper: Crop Yield Response to Water, Rome: Food and Agriculture Organization of the United Nations (FAO), 2012.

USGS (United States Geological Survey): Mineral Commodity Summary Phosphate Rock, 2007.

Van Vuuren, D. P., Bouwman, A. F., and Beusen, A. H. W.: Phosphorus demand for the 1970–2100 period: A scenario analysis of resource depletion, Glob. Environ. Change, 20, 428–439. https://doi.org/10.1016/j.gloenvcha.2010.04.004, 2010.

Proc. IAHS, 376, 87–95, 2018
https://doi.org/10.5194/piahs-376-87-2018

Open Access
PIAHS

Land susceptibility to soil erosion in Orashi Catchment, Nnewi South, Anambra State, Nigeria

Shakirudeen Odunuga, Abiodun Ajijola, Nkechi Igwetu, and Olubunmi Adegun

Department of Geography, University of Lagos, Lagos, Nigeria

Correspondence: Shakirudeen Odunuga (sodunuga@unilag.edu.ng)

Received: 14 June 2017 – Accepted: 4 August 2017 – Published: 1 February 2018

Abstract. Soil erosion is one of the most critical environmental hazards that causes land degradation and water quality challenges. Specifically, this phenomenon has been linked, among other problems, to river sedimentation, groundwater pollution and flooding. This paper assesses the susceptibility of Orashi River Basin (ORB) to soil erosion for the purpose of erosion control measures. Located in the South Eastern part of Nigeria, the ORB which covers approximately 413.61 km^2 is currently experiencing one of the fastest population growth rate in the region. Analysis of the soil erosion susceptibility of the basin was based on four factors including; rainfall, Land use/Land cover change (LULC), slope and soil erodibility factor (k). The rainfall was assumed to be a constant and independent variable, slope and soil types were categorised into ten (10) classes each while the landuse was categorised into five classes. Weight was assigned to the classes based on the degree of susceptibility to erosion. An overlay of the four variables in a GIS environment was used to produce the basin susceptibility to soil erosion. This was based on the weight index of each factors. The LULC analysis revealed that built-up land use increased from 26.49 km^2 (6.4 %) in year 1980 to 79.24 km^2 (19.16 %) in 2015 at an average growth rate of 1.51 km^2 per annum while the light forest decreased from 336.41 km^2 (81.33 %) in 1980 to 280.82 km^2 (67.89 %) in 2015 at an average rate 1.59 km^2 per annum. The light forest was adjudged to have the highest land cover soil erosion susceptibility. The steepest slope ranges between 70 and 82° (14.34 % of the total land area) and was adjudged to have the highest soil susceptibility to erosion. The total area covered of the loamy soil is 112.37 km^2 (27.07 %) with erodibility of 0.7. In all, the overlay of all the variables revealed that 106.66 km^2 (25.70 %) and 164.80 km^2 (39.7 %) of the basin has a high and very high susceptibility to soil erosion. The over 50 % high susceptibility of catchment has serious negative implications on the surface water in terms of water quality and downstream siltation with great consequences on biodiversity and ecosystem services including domestic and industrial usage.

1 Introduction/background

Soil erosion is one of the most critical environmental hazards that causes land degradation and water quality challenges. Several lands had been abandoned in most of the eastern states of Nigeria including Nnewi South owing to a combination of reasons championed by soil erosion. This singular environmental problem is the most prominent hazard in most eastern states of Nigeria where for instance more than 10 % of land resources in Nnewi South is already lost to erosion (Nwajide and Hoque, 1979). Soil erosion is exacerbated due to the decrease in the erosional resistance of the land surface or there is an increase in the erosivity factors acting on the land (Ofomata, 2000). The lost land resources is at the detriment of the local people (peasant farmers) who depend on them for sustainability and livelihood. But these set of people may have ignorantly created or aided the erosion by landuse practice which perhaps was based on their level of understanding. The actions of the highly erosive rainfall energy on the land as well as the topography and soil characteristics with the interferences of humans continues to scar the land, and the effects are cumulatively seen in the sedimentation of rivers, land degradation, pollution of groundwater, land subsidence, flooding etc. However, the process of

Published by Copernicus Publications on behalf of the International Association of Hydrological Sciences.

soil erosion and sediment transport are strongly influenced by landuse changes.

Land use describes the various ways in which humans make use of and manage the land as a resource. It is characterized by the arrangement; activities and inputs people undertake in a certain land cover type to produce, change or maintain it (FAO/UNEP, 1999). A more inclusive definition of land use is often used in practice. The earth surface is shaped by a combination of physical processes, including earthquakes and volcanoes, shifting rocks and sediments, and flows of water and ice. Humans also shape the land through increasing populations and agricultural expansion, mineral, and forest resource excavation, changing the flow of rivers and with layers of industrial and urban infrastructures. Viewed from a holistic perspective, land use concerns the products and/or benefits obtained from the use of land as well as the land management actions carried out by humans to produce those products and benefits.

Land use and land management practices have a major impact on the natural resources including water, soil, nutrient, plants and animals. In addition to the soil erosion, other significant effect of land use includes urban sprawl, land degradation, salinization, and densification. Land degradation has been exacerbated where there has been an absence of any land use planning or one-sided central planning. This has further led to the over-utilization of land resources, for instance, for immediate production at all costs. Consequently, the result has often been a misery for large segment of local population and destruction of valuable ecosystem (Karamouz et al., 2006). Such narrow approaches should be replaced by techniques incorporating planning and management of land resources in an integrated way that ensure the long-term quality of land for human use. Unsustainable land use practices culminating in environmental challenge of soil erosion are often aided by anthropogenic actions earlier mentioned on land, particularly in developing countries.

Rather than battling with the menace of soil erosion, preventing unsustainable and or unfriendly environmental activities on fragile land that could lead to this challenge would be more appropriate particularly in areas of fragile soils such as mid or eastern Nigeria.

It is pertinent also to ask if land resistant or susceptibility to soil erosion is considered before any activity on them is embarked on. The inability to consider this vital question would be synonymous to deliberate attempt on environmental degradation of soil erosion. Soils within the ORB are fertile (which encourages agricultural practices) but also highly erodible due to lack of conservation practice. The susceptibility of the ORB – a region currently experiencing one of the fastest population growth rates in the south-eastern Nigeria requires an erosion control measures which formed the basis of this paper. Soil erosion in the scarp land of the south-eastern Nigeria is generally attributed to man's interference with nature's balance in the soil in addition to the application of destructive methods of land cultivation such as making ridges along slopes (Ofomata, 1985; Oyegun, 1990).

2 Study area

Orashi River Basin (ORB) – a tributary of the River Niger drains Orlu, Orsu, Ideato north and south local government areas of Imo state and Ihiala, Aguata, Ekwusigo, and Nnewi north and south of Anambra state, south-east Nigeria (Fig. 1a). Its geographic limit extends from longitude 6°49′18″ to 7°9′29″ and latitude 5°45′39″ to 50°59′41″. The study area falls within the tropical rainforest vegetative zone; dominant type of trees are oil palm (*Elaeisguineensis*) and mahogany (*Swieteniamacrophylla*) meanwhile shrubs and trees often build-up to thick forest along the Orashi river banks (Fig. 1b). Humid tropical climate with annual rainfall from 1750 to 2500 mm characterized the catchment while the soil types are lateritic in nature, derived from the underlying sandstone and shale units. The soils derived from shale are on the lower slope where drainage is poor, although the occurrences of loamy and humus soils have enhanced the cultivation of crops. Nnewi, Ukporaku, Ubu-Ihembosi, Idemili, Ubu-Otih, Ebubedike, Uchu, and Aghomili streams are the tributaries of the catchment. Orashi basin is an emerging and leading centre for entrepreneurial activities, development of commerce and industry among the people of the south-eastern states of Nigeria. Other economic activities include agriculture, hunting, trading, fishing, craft making, and cottage industries. However, farming activities is dominated by subsistence shifting cultivation where vegetables, corn, cassava, yam, melon and cocoyam are grown.

3 Materials and methods

Data types, characteristics, and sources are shown in Table 1. Based on expert judgment (Tsangaratos and Rozos, 2013), field knowledge and experience, land susceptibility to soil erosion was analysed using the Multi criteria decision analysis (MCDA) (Tsangaratos and Rozos, 2013). This is done by scoring three environmental parameters (slope, soil and land use) based on pre-defined rating. The rating for the class categories (slope, soil and land use) was based on the relative importance of the class or group with respect to soil erosion. The study area was earlier divided into regular grid cells (raster) of 400 m^2 each. In addition, the operational susceptibility score of individual parameter (slope, soil and land use) for each cell is taken as the score for the class with the highest coverage within the cell area. In other words, the cell form the mapping unit for land susceptibility to erosion.

3.1 Rainfall data and landuse/landcover

Climatic characteristics affect soil loss/sediment yield within a basin (Suresh, 2008). Rainfall is an important factor affecting erosion owing to its power to detach the soil particles and

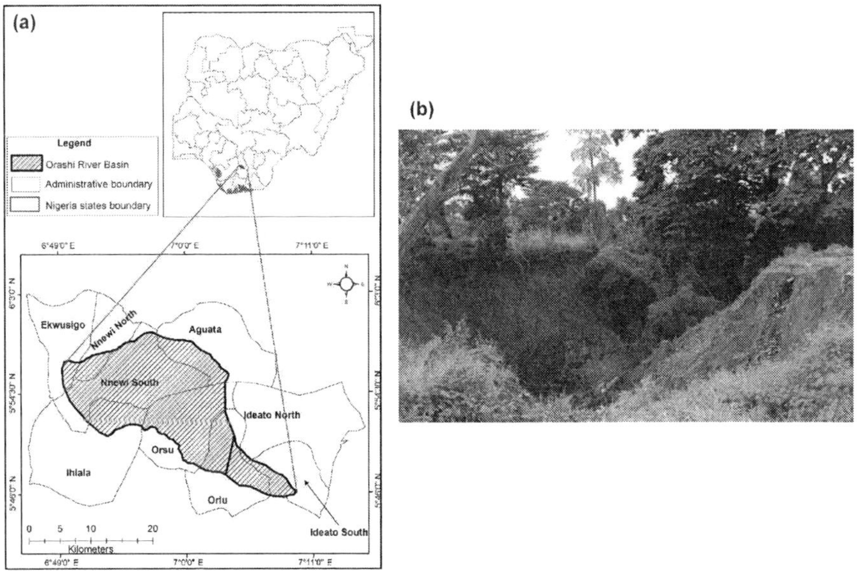

Figure 1. (**a**) Location of study area, Nigeria map inset. (**b**) View of eroded land within the study area.

Table 1. Data sources, characteristics and processes.

Data	Characteristics	Scale	Source
Topographic map	Sheet 312 SE (1964) Sheet 313 NW (1964)	1 : 50 000 1 : 50 000	Nigerian Federal Survey
Satellite imagery	Landsat 4: Bands 7, 4, 2 (1980) Landsat 7: Bands 7, 4, 2 (2000) Landsat 8: Bands 7, 5, 3 (2015)	30 m 30 m 30 m	Earth explorer (https://earthexplorer.usgs.gov)
Digital elevation	Nigeria DEM	10 m	
Soil map	Nigeria Soil Map	state-level	
Climate: rainfall	Enugu Weather Station Owerri Weather Station	3 monthly average	Nigerian Meteorological Agency

Source: authors (2017).

its ability to produce runoff, which causes erosion and transportation of the eroded material. Rainfall was held constant in our analysis so that other variables would form the basis of our assessment of erosion susceptibility across the basin. As shown in Table 1, the 2015 data was obtained from two nearby weather stations of Enugu and Owerri and were used to interpolate the spatial pattern of the rainfall through the categorization of weather records into four (4) periods based on the wet and dry seasons (DJF, MAM, JJA, and SON). The average aggregate of each of these periods was used to determine the rainfall pattern. The wettest period was subsequently used to determine the susceptibility of the basin to soil erosion.

The USGS landuse classification scheme was adopted to classify the various types of landuse/landcover classes on the Landsat imageries. The classification was used for the three periods (scenarios) under investigation (1980, 2000 and 2015). Training dataset was randomly sampled from the spectral signature of each of the imagery to define their respective landuse/landcover type. Five primary classes extracted from the imageries are shown in Table 2. The change detection analysis was performed. The ENVI analytical tool was used to extract the image difference between two periods of time. The five (5) identified land use classes were assigned susceptibility values of 1 to 5 (ranking) where 1 is least susceptible and 5 is highly susceptible land use (Table 2) based on expert judgment.

Table 2. Classification of the landuse at Orashi River Basin.

Code	Class	Susceptibility
1	Built-up area	4
2	Water body	1
3	Heavy forest	2
4	Light forest	5
5	Wetland/marsh land	3

Source: authors (2017).

Table 3. Soil erodibility factor *k*.

Soil	Range of *k*
Loams	0.4–0.7
Clay loams	0.3–0.4
Sandy loams	0.1–0.3
Gravely loams	0.03–0.1

Source: adopted from Garde and Kathyari (1990).

3.1.1 Slope and soil erodibility factor *k*

Basin hydrology and flow characteristics by terrain morphologies of aspect, slope, hill shading, elevation, and curvature and drainage pattern were identified using the "Surface Analysis" toolbox. The raster DEM pixels of each of these parameters were extracted using the surface algorithms to quantify and characterize the morphologic parameter of the basin. Other quantitative variables determined within study area include bifurcation ratio, stream length, stream frequency, circulatory ratio.

Erosion increases with the increase in slope and length of the slope (Garde and Kathyari, 1990); this is due to corresponding increase in the velocity and volume of surface runoff. The slope of the basin was classified into ranges and further reclassified according to their degree of the susceptibility as shown in Table 5. Soil erodibility depends on the texture, aggregate stability, shear strength, infiltration capacity, and organic and chemical contents. The coarser the particles the smaller the value of their erodibility. Similarly, the greater the relative density the less the erodibility (Garde and Kathyari, 1990). Table 3 shows soil erodibility Factor adopted from Garde and Kathyari (1990), while Table 6 shows the erosion susceptibility classification of soil erodibility factor used for this study.

3.1.2 Factor of soil erosion susceptibility integration

Rainfall parameter was held as a constant among the other variables considered for the susceptibility of the basin to erosion. The slope of the basin was classified into ranges and further reclassified according to their degree of susceptibility while the landuse parameter was obtained from the current landuse/landcover pattern. Each of the classes of the landuse was reclassified based on the rate of the sediment generated. The rating of the classes within each parameter was based on the relative importance of each class obtained from field experience and expert knowledge. However, the maximum susceptibility weight/score obtainable for land use was 5, slope was 10 and soil erodibility factor was also10 for each cell. This was based on the fact that both slope and soil erodibility factors are predominantly the physical attributes of the environment while land use on the other hand is human induced catchment management. Any alteration (natural or human) on the physical attribute of the terrain can trigger erosion while landuse can be managed and controlled not to induce erosion. Also, landuse activities that cause erosion do so by impacting either slope or soil erodibility factor. It is based on this domineering impact of slope and erodibility factor in erosion, especially gullying that the score range of slope and erodibility factor were twice that of the land use. The influence of each parameter on the erosion has already been weighted through the expert judgments class scores. The scores were reclasified to a uniform susceptibility rating scale through a combination of the scores at the cell level (i.e. the total susceptibility is the sum of the ranking scores for each component at each grid). Five (5) classes (Table 8) of soil susceptibility (very low susceptibility, low susceptibility, medium susceptibility, high susceptibility and very high susceptibility) were generated and used for mapping. This linear combination approach is one of the best known and most commonly used MCDA methods (Malczewski, 1999; Ayalew et al., 2004; Tsangaratos and Rozos, 2013) for susceptibility analysis. The summing ranking-based scores at cell unit provide the level of exposure of the unit area taken into consideration all parameters leading to erosion at this micro-unit while maintaining the variation in the spatial interplay of the parameters at the macro-level.

4 Result and discussion

4.1 Analysis of Orashi River Basin landuse/landcover change trajectory

Table 4 shows the landuse/landcover dynamics of the ORB while Figs. 2–4 graphical display the LULC pattern. Five classes of land uses; built-up (area), water body, wetland, light and heavy forest vegetations were identified. The Dominant light forest vegetation at the base year had witnessed a continuous decrease (Ochege and Okpala-Okaka, 2017) from 336.41 km^2 (81.33 %) to 303.71 km^2 (73.43 %) and 280.82 km^2 (67.89 %) in 2000 and 2015 respectively. Massive decrease occurred between 1980 and 2000 (2 decades) with an annual decline of about 1.65 km^2. Between 2000 and 2015, about 23 km^2 was lost with an annual decrease of 1.53 km^2. The decrease in the forest vegetation is strongly connected to the increase in the population of the study basin, hence, the high demand for residential and arable lands (Lal et al., 1992) as reflected in the built-up areas which increased

Table 4. Landuse/landcover change.

Landuse	1980		2000		2015	
	Area (km^2)	%	Area (km^2)	%	Area (km^2)	%
Built-up	26.49	6.4	48.66	11.77	79.24	19.16
Heavy forest	22.51	5.44	22.28	5.39	13.53	3.27
Light forest	336.41	81.33	303.71	73.43	280.82	67.89
Water body	0.17	0.04	0.25	0.06	0.04	0.01
Wetland	28.03	6.78	38.7	9.36	39.98	9.67
Total	413.61	100	413.61	100	413.61	100

Source: authors (2017).

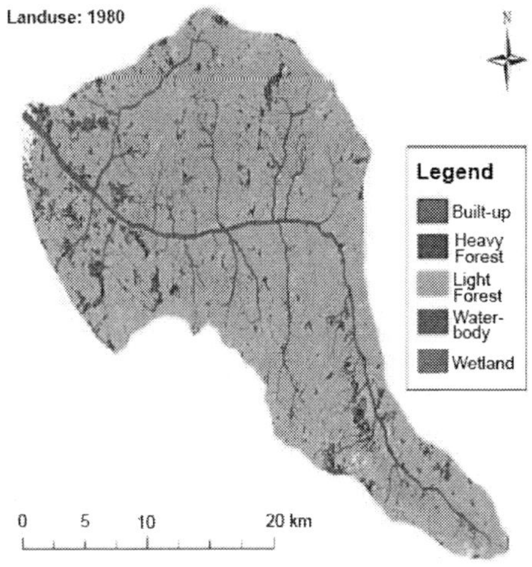

Figure 2. Landuse/land cover pattern in 1980.

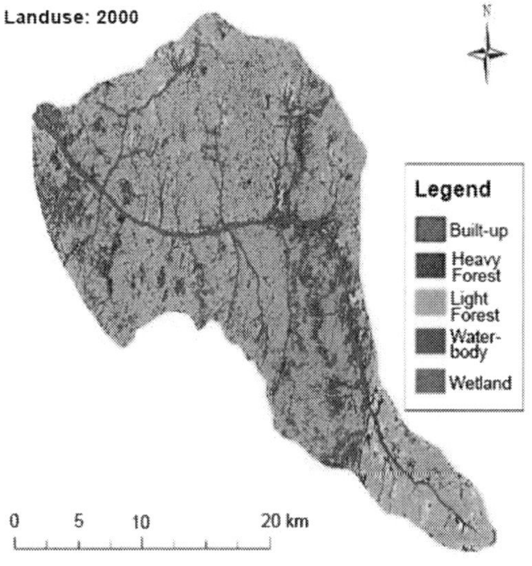

Figure 3. Landuse/landcover pattern in 2000.

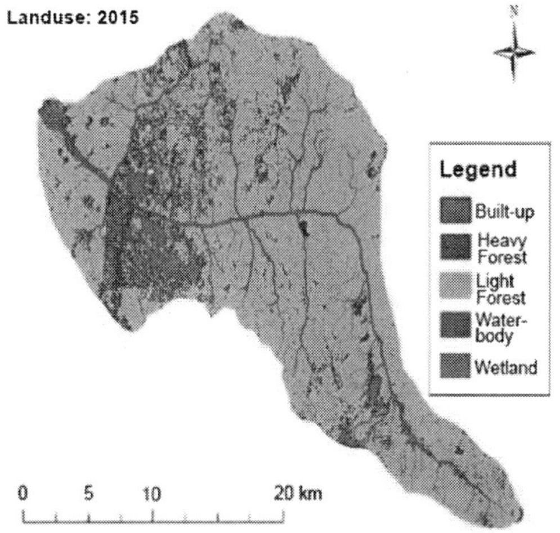

Figure 4. Landuse/landcover pattern in 2015.

in a trajectory mode from 26.49 km^2 in 1980 to 79.24 km^2 in 2015. Within 35 years (1980–2015), the light forest vegetation decreased from 81.33 to 67.89 % of the total land area of ORB.

Conversely, the increase in the wetland from 6.78 % (1980) to 9.67 % (2015) could be linked to bank erosion where areas that were either light forest or heavy forest are eroded to become wetland. Also, river siltation increases wetland area by depositing sediments from the upland in the river course, thereby, reducing the water body to the wetland area. In other words, the increase in the expanse of the wetland must have been aided by floodplain deposition of the eroded soil due to the reduced velocity that decreases the transportation potency of the rivers. The increase in the built-up lands from 26.29 km^2 (6.4 %) in 1980 to 79.24 km^2 (19.2 %) in 2015 at an annual growth rate of 1.51 km^2 shows an evidence increase in population (Igbokwe et al., 2013). Significant and highest annual growth rate of 2.04 km^2 occurs between 2000 and 2015 without predesign layout or any form of developmental plan whatsoever. This, however, has increased the contribution of anthropogenic activities especially the built-up landuse to soil erosion in the environment.

The insignificant change in the heavy forest vegetation (0.29 %) for the thirty years period could be linked to tradition and culture of the people of Nnewi, which hold thick forests as sacred where spiritual artifacts (shrines) reside and where other religious or traditional activities take place. However, between 2000 and 2015, a drastic change was observed where the heavy forest vegetation decreased (Ochege and Okpala-Okaka, 2017) from 22.28 to 13.53 km^2 at an annual decrease of 0.58 km^2, the recent pressure on the heavy forest may be owing to the population increase and degraded

Table 5. Slope erosion susceptibility classification.

Slope	Grid count	% of area covered	Susceptibility
0–5	173 347	16.76	1
5–12	102 181	9.88	2
12–21	68 841	6.66	3
21–30	75 057	7.26	4
30–40	64 429	6.23	5
40–50	113 489	10.97	6
50–57	100 275	9.7	7
57–63	67 727	6.55	8
63–70	120 741	11.67	9
70–82	148 131	14.32	10
	1 034 218	100	

Source: authors (2017).

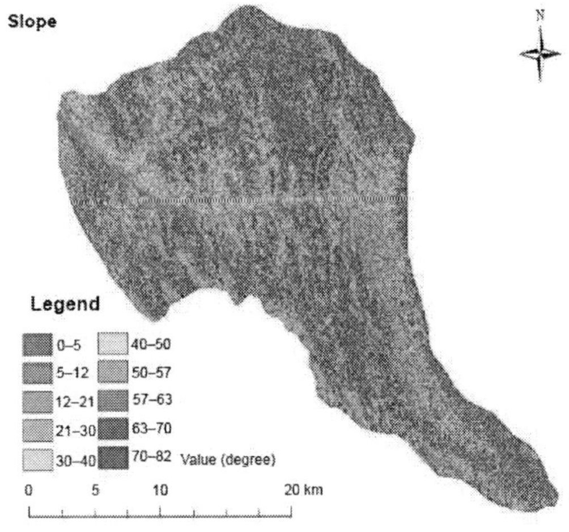

Figure 5. Slope morphology of the basin.

soil of the light forest vegetation, hence, the need for more agricultural fertile/productive lands.

The dominant landcover within the study area is the light forest vegetation, however, it is highly susceptible to further depletion and degradation with corresponding surface soil loss of the basin.

4.1.1 Morphological characteristics of Orashi River Basin

The slope describes the rate of river dissection within the basin and also determines the rate of soil erosion. In other words, soil erosion increases as the slope increases. Table 6 and Fig. 5 show the slope characteristics of the basin as derived from the grid codes percentage. Three categories of the erosional processes were found according to this classification. Slopes within 70–82° (14.32 %), 63–70° (11.67 %), 57–63°, (6.55 %), 50–57° (9.70 %) and 40–50° (10.97 %) are categorized as the highest slope. This indicates a slope of active erosional processes. The second category is classified as a slope of medium erosional processes, which ranges 30–40° (6.23 %), 21–30° (7.26 %) and 12–21° (6.66 %). The last category describes a slope of low erosional processes. This slope is relatively flat and found in area of deposition such as river mouth and river channel. The slope ranges between 5–12° (9.88 %) and 0–5° (16.76 %).

Table 6. Soil classification and erodibility of Orashi.

Soil class	Area	%	Erodibility	Susceptibility
Sand	5.65	1.36	0.09	1
Gravely loams	24.31	5.86	0.1	2
Loamy clay	30.79	7.42	0.2	3
Sandy clay	1.02	0.25	0.27	4
Sandy loam	56.29	13.56	0.3	5
Loamy sand	2.33	0.56	0.38	6
Clay loam	21.14	5.09	0.4	7
Clay	53.52	12.89	0.45	8
Sandy clay loam	107.69	25.94	0.48	9
Loam	112.37	27.07	0.7	10
Total	415.1	100		

Source: authors (2017).

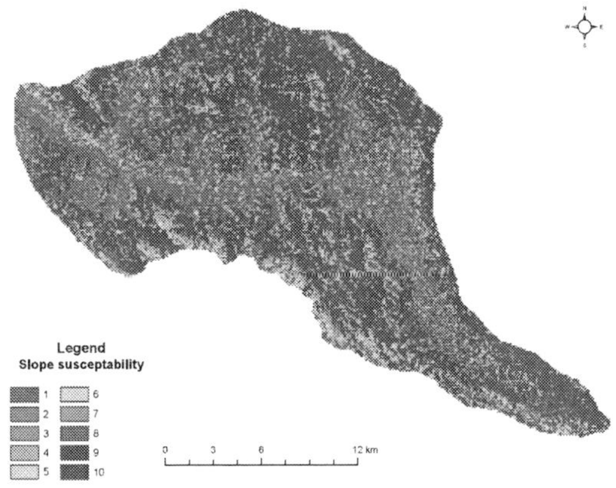

Figure 6. Slope susceptibility.

4.1.2 Susceptibility of Orashi Basin to soil erosion

Slope susceptibility

Orashi basin slope classifications and their level of susceptibility are shown in Table 5. Ten (10) classes of susceptibility levels were identified. Slope within 0–5° which covers about 16.76 % of the basin has a level 1 value (Fig. 6). This indicates that at this level erosion activities is minimal, site specific and low impact while at slope within 70–

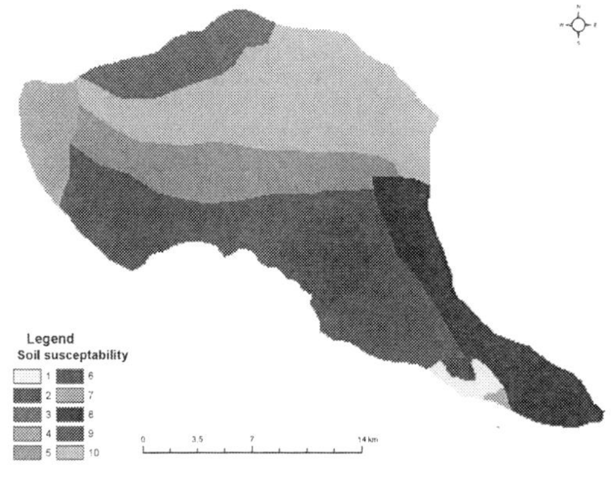

Figure 7. Soil susceptibility classification of Orashi Basin.

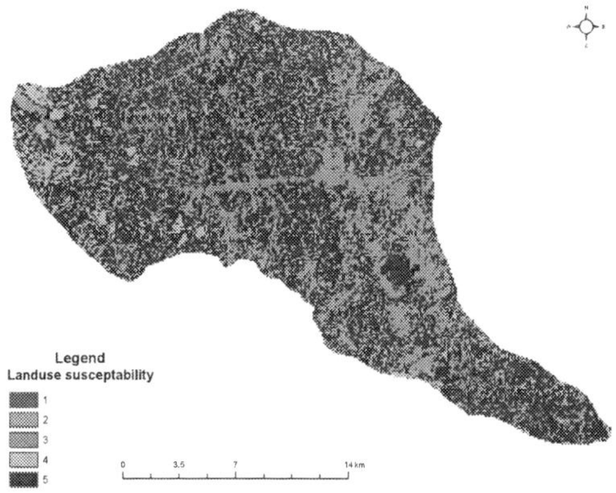

Figure 8. Landuse susceptibility classification.

Table 7. Landuse erosion susceptibility classification.

Landuse	Area (km^2)	%
Waterbody	0.04	0.01
Heavy forest	13.53	3.27
Wetland	39.98	9.67
Built-up	79.24	19.16
Light forest	280.82	67.89
Total	413.61	100

Source: authors (2017).

Table 8. Basin areas susceptible to erosion.

Sn	Total susceptibility score	Class	Area
1	0–5	Very low susceptibility	57.84
2	6–10	Low susceptibility	45.48
3	11–15	Medium susceptibility	40.31
4	16–20	High susceptibility	106.66
5	21–25	Very high susceptibility	164.8
Total			415.09

Source: authors (2017).

82°, about 14.32 % falls within highest erosion susceptibility level (Akbarimehr and Jalilvand, 2013; Zhang et al., 2015) of Level 10. This indicates that the rate of soil erosion is very intensive and a lot of top soils are freely removed as shown in Fig. 6.

Soil erodibility and susceptibility

The soil types within the basin were dominated by sandy clay loam (25.94 %) and loam (27.07 %) Table 6. In other words, the basin is layered by fertile soil good for agricultural activities. These dominant soil classes (sandy clay loam and loam) have erodibility of 0.48 and 0.7 respectively with high susceptibility of 9 and 10 levels respectively. These implies a high fragility to soil erosion. Figure 7 shows the soil susceptibility classification of Orashi Basin.

4.1.3 Landuse susceptibility

Table 7 shows the landuse of the basin with their degrees of susceptibilities; light forest (280.80 km^2, 67.89 %) has the highest susceptibility value of level 5. This is followed by built-up area (79.24 km^2, 19.16 %) which has a susceptibility value of level 4. Other LULC are wetland (level 3), heavy forest (level 2), and water body (level 1) with their respective coverage of 39.98, 13.53, and 0.04 km^2 respectively. Figure 8 shows the landuse susceptibility classification.

4.1.4 Susceptibility of the basin to soil erosion

The overall susceptibility of the basin through variables integration using overlay analysis in GIS environment was performed by converting variables scores in the grid (400 m^2 raster format) and the Math Raster Calculator is used to sum up the aggregate score for each cell and to assign the cell to soil erosion susceptibility class areas as shown in Table 8. The result reveals that about 164.8 km^2 (39.7 %) of the ORB is highly susceptibility to soil erosion. These susceptibility levels are: very low (5–9), low (10–13), medium (14–15), high (16–19) and very high (20–25) Table 8 and Fig. 9. In each of these classes, however, Ihemeposi community falls within very low susceptibility, Ozu and Okohia communities fall within low susceptibility, Upor and Azra (medium susceptibility), Ozubulu and Uruobo Okija high susceptibility while Otu, Osu, Ezinifite and Akwa communities falls under very high susceptibility. Consequent upon this, these variables were converted into grid (raster format: 400 m^2) and

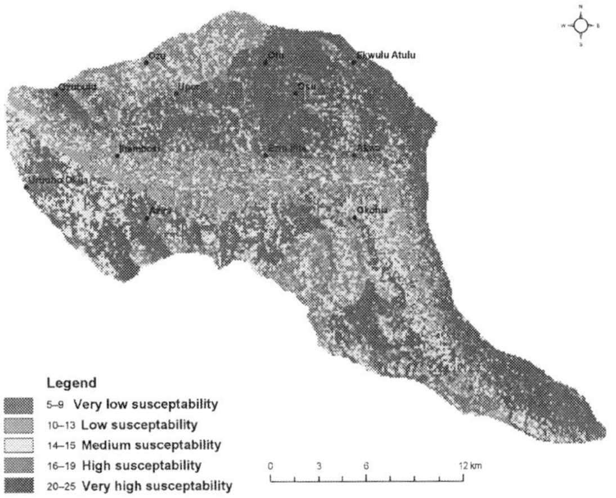

Figure 9. Areas susceptible to erosion.

combine together using Math Raster Calculator to determine the areas that are susceptible to soil erosion.

5 Conclusion

The study reveals that a substantive expanse of the ORB area is susceptible to soil erosion and land degradation. The over 50 % catchment area that falls under high susceptibility has serious negative implications on the land utilization and surface water uses in terms of water quality and downstream siltation. Also, these have great consequences on the biodiversity and ecosystem services, including the domestic use of water and subsistence farming. However, an increase in population and its ancillary developmental activities requires more land area for cultivation so as to ensure sustainable food supply and subsistence livelihoods. If these needs are to be met in a sustainable way, there would be a need to change the present land use practices to a more integrated system that would relate all activities to the basin development. It is also recommended that public awareness on soil/land conservation, strategic planning and land management be embarked upon.

Data availability. Links to the USGS Satellite Imagery and Digital Elevation Model (DEM) data are available at Earth explorer (https://earthexplorer.usgs.gov). Other data such as soil, weather, and topographic maps could not be accessed publicly or online (website) for reasons best known to the custodians of these data in Nigeria.

Author contributions. SO conceptualized and led the team on this research. In addition to the leadership and technical roles he played, he also designed the paper layout. AA prepared the manuscript and effected corrections made during the course of the write-up. NI gathered and processed all the spatial data used for this article. OA reviewed the article from time to time for necessary corrections and also serves a "liaison" officer between the research team and Copernicus (IAHS).

Competing interests. The authors declare that they have no conflict of interest.

Special issue statement. This article is part of the special issue "Water security and the food–water–energy nexus: drivers, responses and feedbacks at local to global scales". It is a result of the IAHS Scientific Assembly 2017, Port Elizabeth, South Africa, 10–14 July 2017.

Acknowledgements. Authors wish to express their profound gratitude to the Office of Surveyor General of the Federation (OSGOF) and Nigeria Meteorological Agency (NIMET) for providing the topographical maps and weather data respectively. Our thanks also go to the Department of Geography, University of Lagos for providing enabling environment which facilitated putting this work together. The assistance and support privileged by the staff of Nnewi South Local Government Area of Anambra State during the field data gathering is also acknowledged.

Edited by: Barry Croke
Reviewed by: two anonymous referees

References

Akbarimehr, M. and Jalilvand, H.: Considering the relationship of slope and soil loss on skid trails in the north of Iran (a case study), Journal of Forest Science, 59, 339–344, 2013.

Ayalew, L., Yamagishi, H., and Ugawa, N.: Landslide susceptibility mapping using GIS-based weighted linear combination, the case in Tsugawa area of Agano River, Niigata Prefecture, Japan, Landslides, 1, 73–81, 2004.

FAO/UNEP: The Future of Our Land, Guidelines for Integrated Planning for Sustainable Management of Land Resources, FAO (Rome), UNEP (Nairobi), 1999.

Garde, R. J. and Kathyari, U. C.: Erosion prediction models for large catchments, in: Proceedings of the International Symposium on Water Erosion, Sedimentation, and Resource Conservation, Dehradun, India, 89–102, 1990.

Igbokwe, J. I., Ezeomedo, I. C., and Ejikeme, J. C.: Identification of urban sprawl using GIS Technique: A case study of Onisha and its environs in south east, Nigeria, International Journal of Remote Sensing & Geoscience, 2, 41–49, 2013.

Karamouz, M., Hashemi Olya, R., Moridi, A., and Ahmadi, A.: Water Allocation from River System Considering the Interaction between River and Aquifer, ASCE Environmental and Water Resources Institute Conference, Delhi, India, 2006.

Lal, R., Ghuman, B. S., and Shearer, W. A.: Cropping systems effects of a newly cleared Ultisol in Southern Nigeria, Soil Technol., 5, 27–38, 1992.

Malczewski, J.: GIS and Multicriteria Decision Analysis, John Wiley and Sons, New York, 1999.

Nwajide, S. C. and Hoque, M.: Gullying processes in south-eastern Nigeria, The Nigerian Field, 44, 64–74, 1979.

Ochege, F. U. and Okpala-Okaka, C: Remote Sensing of vegetation cover changes in the humid tropical rainforest of southeastern Nigeria (1984–2014), Cogent Geoscience, 3, 1307566, https://doi.org/10.1080/23312041.2017.1307566, 2017

Ofomata, G. E. K.: Soil Erosion in Nigeria: The Views of a Geomorphologist, University of Nigeria, Nsukka, Nigeria, Inaugural Lecture Series No. 7, 3–33, 1985.

Ofomata, G. E. K.: Classification of Soil Erosion with Specific reference to Anambra State of Nigeria, Environ. Rev., 3, 252–2551, 2000.

Oyegun, C. U.: The management of coastal zone erosion in Nigeria, Ocean and Shoreline Management, 14, 215–228, https://doi.org/10.1016/0951-8312(90)90035-G, 1990.

Suresh, R.: Watershed Hydrology, Standard publishers, New Delhi, India, 2008.

Tsangaratos, P. and Rozos, D: Producing landslide Susceptibility maps by applying expert knowledge in a GIS – based environment, Bulletin of the Geological Society of Greece, XLVII, 2013.

Zhang, Z., Sheng, L., Yang, J., Chen, X.-A., Kong, L., and Wagan, B.: Effects of Land Use and Slope Gradient on Soil Erosion in a Red Soil Hilly Watershed of Southern China, Sustainability, 7, 14309–14325, https://doi.org/10.3390/su71014309, 2015.

Proc. IAHS, 376, 97–104, 2018
https://doi.org/10.5194/piahs-376-97-2018

Potential impact of climate change to the future streamflow of Yellow River Basin based on CMIP5 data

Xiaoli Yang[1,2], Weifei Zheng[1,2], Liliang Ren[1,2], Mengru Zhang[1,2], Yuqian Wang[1,2], Yi Liu[1,2], Fei Yuan[1,2], and Shanhu Jiang[1,2]

[1]State Key Laboratory of Hydrology-Water Resources and Hydraulic Engineering, Hohai University, Nanjing, China

[2]College of Hydrology and Water Resources, Hohai University, Nanjing, China

Correspondence: Liliang Ren (rll@hhu.edu.cn)

Received: 6 June 2017 – Accepted: 22 August 2017 – Published: 1 February 2018

Abstract. The Yellow River Basin (YRB) is the largest river basin in northern China, which has suffering water scarcity and drought hazard for many years. Therefore, assessments the potential impacts of climate change on the future streamflow in this basin is very important for local policy and planning on food security. In this study, based on the observations of 101 meteorological stations in YRB, equidistant CDF matching (EDCDFm) statistical downscaling approach was applied to eight climate models under two emissions scenarios (RCP4.5 and RCP8.5) from phase five of the Coupled Model Intercomparison Project (CMIP5). Variable infiltration capacity (VIC) model with 0.25° × 0.25° spatial resolution was developed based on downscaled fields for simulating streamflow in the future period over YRB. The results show that with the global warming trend, the annual streamflow will reduced about 10 % during the period of 2021–2050, compared to the base period of 1961–1990 in YRB. There should be suitable water resources planning to meet the demands of growing populations and future climate changing in this region.

1 Introduction

Climate change through the temperature, precipitation, evaporation and other factors to change the impact of hydrological circulation system, leading to different spatial and temporal scale of water redistribution (Stocker, 2013).With the global climate changes, both the precipitation and runoff in Yellow River Basin (YRB) present obviously decreasing, while the water shortage of the Yellow River basin will likely be intensifying with the economic development (Wang et al., 2011; Hao et al., 2011; Duan et al., 2014). Meanwhile, YRB is well known not only for its history and large drainage area but also for its frequent floods and serious droughts. So, it is important to analysis climate change and its impact on the hydrological process in YRB.

The global climate models (GCMs) driving hydrological models to calculate watershed hydrological conditions is a common means of studying the effects of climate change on water resources (Zhang and Wang, 2007). However, the coarse resolution makes GCMs' simulation of regional and extreme weather events accuracy is not high (Xu et al., 2010). Downscaling technology is a bridge that connects GCMs low-resolution output to high-resolution meteorological elements for hydrological model (Wilby et al., 2002). Statistical downscaling methods are widely used for easy to construct, diverse methods, and flexible form (Tareghian and Rasmussen, 2013; Murphy, 2000). Li et al. (2010) proposed an equidistant cumulative distribution function matching method (EDCDFm) based on the quantile function method, which consider the differences between the projected climatic factors and the historical statistical cumulative distribution of climatic factors. This method can effectively capture the extreme elements of climate factors, improving the simulation accuracy of climate factors (Wang and Chen, 2014; Aloysius et al., 2016). In this study, the EDCDFm was used to downscale eight models published in the phase five of the Coupled Model Intercomparison Project (CMIP5) of the GCMs over the YRB from 1961 to 2099 under RCP4.5 and RCP8.5 two emission scenarios. The performances of

Table 1. The information of eight CMIP5 models.

Model Version	Institute	Horizontal Resolution (latitude × longitude)
bcc-csm1-1	Beijing Climate Center	2.8° × 2.8°
CanESM2	Canadian Center for Climate Modeling and Analysis	2.8° × 2.8°
CCSM4	National Center for Atmospheric Research	1.25° × 0.9°
CSIRO-Mk-3-6-0	CSIRO Marine and Atmospheric Research	1.875° × 1.875°
GISS-E2-R	Goddard Institute for Space Studies, New York, NY	2.5° × 2°
MPI-ESM-LR	Max Planck Institute for Meteorology	1.875° × 1.875°
MRI-CGCM3	Meteorological Research Institute	1.125° × 1.125°
NorESM1-M	Norwegian Climate Centre	2.5° × 1.875°

downscaled monthly precipitation and temperature of the climate model were analyzed against to the observation data during the historical period (1961–2005). Based on the observation data and downscaled data, the variable infiltration capacity (VIC) hydrological model with 0.25° × 0.25° grid spatial resolution of the YRB was constructed. The Mann-Kendall trend test was used to analyze the trend of simulated streamflow.

2 Data set and methods

2.1 Study area and data set

As the second longest river in China , YRB is an important agricultural production base in China, which locates in 32–42° N and 94–119° E. YRB has a complex surface, large heterogeneity of surface vegetation, which is sensitive to climate change and human disturbance. The annual temperature in YRB has an obviously spatial uneven, which is higher in the southeast but lower in the northwest. The precipitation mainly occurs in summer rains more, southeast rainy, northwest drought, plain precipitation is more than plateau, mountain is more than basin. The whole basin was divided into ten sub-catchments (Fig. 1), including seven catchments: Tangnaihai (TNH), Lanzhou (LZ), Toudaoguai (TDG), Hejin (HJ), Sanmenxia (SMX) and Huayuankou (HYK) with their control stations in the trunk stream, two tributary catchments Xianyang (XY) and Huaxian (HX) and an interior drainage area (located in northern YRB between Wubao (WB) and Longmen (LM) catchments).

We collected observation data (monthly precipitation and monthly values of mean, minimum, and maximum surface air temperature) of 101 weather stations in the YRB for the period of 1961–2005, from China Meteorological Data Service Center (CMDC) (http://data.cma.cn/). The meteorological stations' data was interpolated to interpolate at a spatial resolution of 0.25° × 0.25° using kriging method and the temperature adjusted for differences in elevation between the two grids taking into account the lapse rate of −0.65 °C for every 100 m increase in elevation. The daily streamflow data

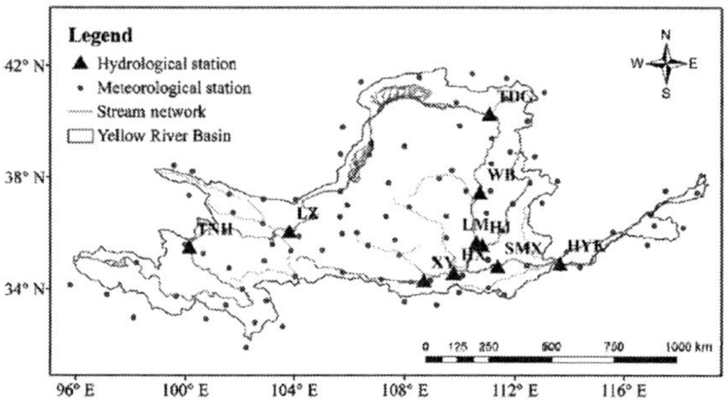

Figure 1. Distribution of 101 meteorological stations and ten hydrological stations in YRB.

of ten hydrological stations were collected from the "China Year Books of Hydrology".

Eight climate models for the climate change projections under the RCP4.5 and RCP8.5 scenarios (Table 1) downloaded from the Program on Climate Model Diagnosis and Intercomparison (PCMDI) website (https://esgf-node.llnl.gov/search/cmip5/). All the historical simulations are for the period of 1961 to 2005 according to the observed data time series, and the future projects of RCP45 and RCP85 cover the period of 2006–2099. Because of the difference of each climate model horizontal resolution, we used bilinear interpolation to make all the GCM climate fields at same spatial resolution (0.25° × 0.25°) with the observations.

2.2 Equidistant quantile-based mapping method

The equidistant CDF matching (EDCDFm) statistical downscaling method developed by Li et al. (2010) was applied to bias correct the monthly precipitation and temperature fields from eight CMIP5 models in YRB. EDCDFm uses the differences between the GCMs simulate climate factors and cumulative distribution characteristics of regional observation climate to deviation correction about the climate, which is simulated by climate models. Compared the standard quantile-based mapping method (CDF), the EDCDFm can capture the

Table 2. The RMSE of monthly precipitation and temperature between observed and downscaled model simulated in YRB from 1961–2005.

	Precipitation (mm)	Mean Temperature (°C)	Max-Temperature (°C)	Min-Temperature (°C)
bcc-csm1-1	37.12	1.82	2.21	1.96
CanESM2	36.39	1.82	2.20	1.95
CCSM4	36.23	1.79	2.18	1.88
CSIRO-MK3-6-0	36.07	1.85	2.35	1.82
GISS-E2-R	36.61	1.88	2.41	1.80
MPI-ESM-LR	36.59	1.90	2.36	1.94
MRI-CGCM3	36.25	1.92	2.31	1.99
NorESM1-M	35.67	1.78	2.13	1.94
Ensemble	28.10	1.40	1.73	1.46

extreme climate effectively, improving internal error climate model data and the limitation of the application of the method of interpolation, and then improve the accuracy of the climate simulation (Eq. 1). In this study, we applied this method first to correct the biases in the historical period (1961–2005) and then to the future period (2006–2099) for each model.

$$x_{\mathrm{m-p,adjst}}=x_{\mathrm{m-p}}+F_{\mathrm{o-c}}^{-1}(F_{\mathrm{m-p}}(x_{\mathrm{m-p}}))-F_{\mathrm{m-c}}^{-1}(F_{\mathrm{m-p}}(x_{\mathrm{m-p}})) \quad (1)$$

where F is the CDF of either the observations (o) or model (m) for a historic training period, current climate period (c) or future projection period (p).

2.3 Evaluation of model performance

Root mean square error (RMSE) and correlation coefficient (CC) were used to evaluate the downscaled model performance against to the observations. The RMSE and CC between a model-simulated and observation climate fields are defined as:

$$\mathrm{RMSE}=\sqrt{\frac{\sum_{i=1}^{n}(M_i-O_i)^2}{n}} \quad (2)$$

and,

$$CC=\frac{\sum_{i=1}^{n}(M_i-\overline{M}_i)(O_i-\overline{O}_i)}{\sqrt{\sum_{i=1}^{n}(M_i-\overline{M}_i)^2\sum_{i=1}^{n}(O_i-\overline{O}_i)^2}} \quad (3)$$

Where M, O, $\overline{M}_i$, $\overline{O}_i$ are the simulated fields, observation fields, and mean of simulated fields, mean of observation fields, respectively. i is the number of spatial grid cells.

2.4 VIC Model

VIC model is a semi distributed, physically based, macroscale surface water and energy balance hydrological model. It can be run in either water balance mode or full water and energy balance mode. The model takes into account the land-gas water balance and energy balance, the unevenness of the spatial distribution of soil water storage capacity, considering streamflow yield under saturated storage and streamflow yield under excess infiltration, nonlinear regression of base flow. The Richards equation is used to describe the vertical one-dimensional soil water movement, and the Darcy law is used to describe the process of water vapor flux in soil layers (Zhang et al., 2009). We used observed data and eight downscaled model's outputs as the climatic forcing data of the VIC model. The VIC model has seven sensitive parameters that need to be calibrated for the study area, including the infiltration curve index (infilt), the lowest maximum base of soil flow (D_{smax}), base flow of nonlinear growth occurs and the proportion of the D_{smax} (D_{s}), the ratio of the bottom soil maximum moisture base flow nonlinear growth (W_{s}), the depth of soil layer (Depth1, Depth2, Depth3). In this study, Nash Sutcliffe coefficient (NSCE; Nash and Sutcliffe, 1970) and total relative Bias were used to evaluate the model simulations.

3 Results and discussions

3.1 Assessment of downscaled outputs

RMSE of monthly precipitation and temperature between downscaled outputs with observations were presented in Table 2. It could be found that the RMSE of eight model's ensemble for both precipitation and temperature is lower than any individual model. Among eight models, model NorESM1-M shows the smallest RMSE between precipitation, mean temperature, maximum temperature, and minimum temperature than any other model. The RMSE of monthly precipitation from individual model varies from 35.67 to 37.12 mm, and model bcc-csm1-1 has the highest RMSE value. The downscaled monthly mean temperature has lowest RMSE varies from 1.78 to 2.41° than that of maximum and minimum temperature. Model MPI-ESM-LR has the highest RMSE of monthly mean temperature, model GISS-E2-R has the highest value of maximum temperature, and model MRI-CGCMS has the highest RMSE value of minimum temperature. The standard deviation and correlation using Taylor figure (Taylor, 2001) of eight models

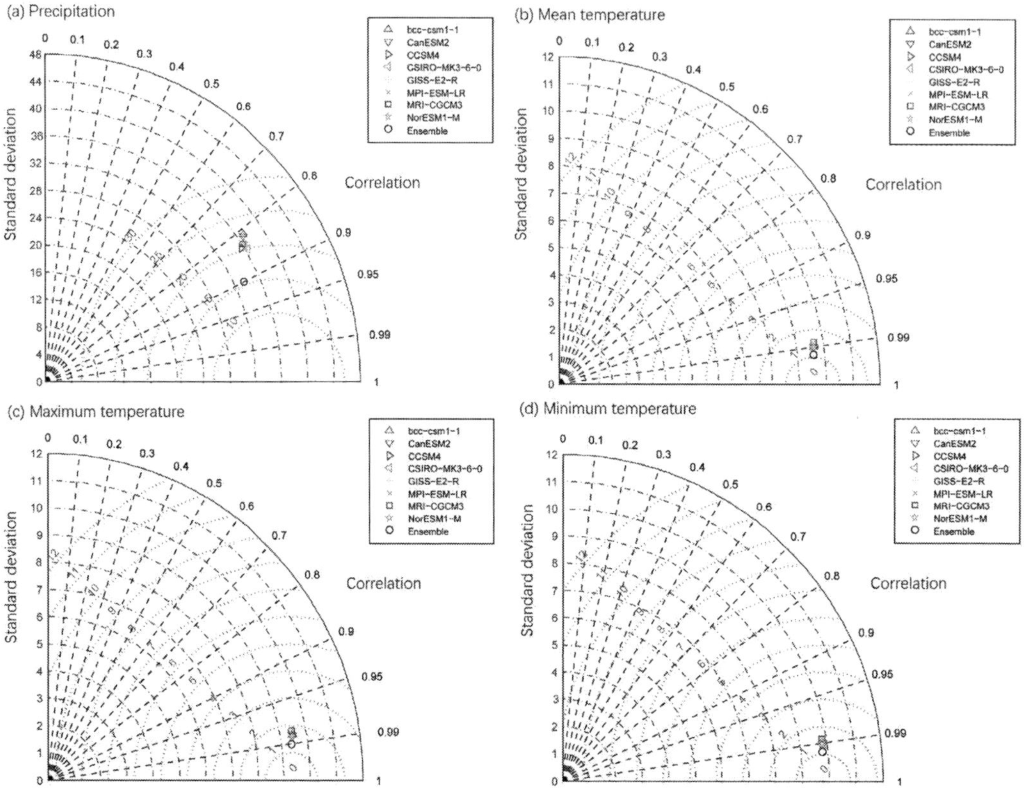

Figure 2. Taylor diagram of standard deviation and correlation of observed and downscaled **(a)** monthly precipitation, **(b)** mean temperature, **(c)** maximum temperature, and **(d)** minimum temperature in YRB (1961–2005).

and their ensemble against to the observations (Fig. 2). The correlation coefficient of precipitation is about 0.83. Meanwhile, the monthly mean, maximum, and minimum temperature have higher correlation coefficient (around 0.99) than that of precipitation. Especially, the multi-model ensembles of precipitation and temperature present higher correlation coefficient than any single model. It shows that the multi-model ensemble can reduce the noisy of each model's simulation. Therefore, we analyzed the spatial distribution of RMSE and CC of eight model simulations against to the observations.

The spatial distribution of RMSE of downscaled ensemble outputs against to the observations was presented in Fig. 3. It shows that the highest RMSE occurred in the east area varies from 56 to 72 mm, while the lowest values occurred in the west area varies from 16 to 24 mm. The RMSE of the maximum and minimum temperature in the middle of north area is larger than other regions. Meanwhile, the west area has the highest correlation coefficient of precipitation varies from 0.8 to 0.9, while middle area has the lowest correlation coefficient with the observed precipitation. The mean and minimum temperature have good correlation coefficient with the observations varies from 0.98 to 0.99 for the whole basin. The correlation coefficient of the maximum temperature is lower than that of mean and maximum temperature, and the west region has the relatively lowest correlation coefficient value (about 0.96). Thus, we can conclude that these downscaled eight model's ensemble is good at simulating the precipitation and temperature in YRB.

3.2 VIC model calibration and validation

We analyzed the NSCE and Bias of the VIC model simulations with the observed streamflow of each hydrological station for calibration period and validation period (Table 3). The whole basin was divided into ten sub-catchments for calibration and validation. For nested catchments, the upstream flood wave was routed into the downstream with the Muskingum method. Then by taking the difference between observed and routed ones of downstream, the hydrograph for areas between upstream and downstream was derived and was further applied for calibration. Since no observation stations

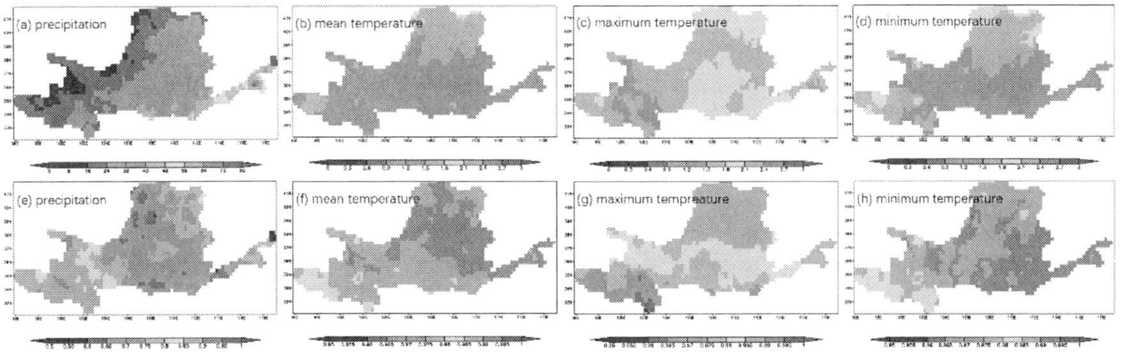

Figure 3. Spatial distribution of RMSE **(a–d)** and CC **(e–h)** showing comparison of monthly model set range simulations over the region with observations (1961–2005).

Table 3. The statistics of calibration and validation periods monthly flows for the ten sub-catchments in YRB.

Hydrology	NSCE		Bias (%)	
Station	Calibration	Validation	Calibration	Validation
TNH	0.879	0.804	−1.6	13.3
LZ	0.833	−0.070	−1.8	7.8
TDG	0.448	−2.73	43.5	86.5
WB	0.979	0.879	3.5	13.8
LM	0.983	0.92	4.3	8.4
SMX	0.951	0.887	4.0	14.7
HYK	0.236	0.394	2.5	9.2
HJ	0.651	0.224	3.1	133.4
XY	0.744	0.176	−1.1	68.9
HX	0.942	0.886	4.6	20.1

were collected for the catchments of interior drainage area and the lower reaches of HYK, parameters from their each adjacent catchment, WB and HYK respectively, were employed for computation instead. The NSCE of seven stations are above 0.7 and the bias of streamflow is around ±5 % during the calibration period (1961–1990). Meanwhile, there are five stations has a higher Nash value above 0.7 during the validation period (1991–2012). These results indicate that VIC model is good at simulating the hydrological procession of the Yellow River.

3.3 Projected changes of precipitation and temperature

During future period of 2021–2050, the annual precipitation for two RCPs show an increase trends, and the increasing rate varies from 7.795 mm/10 a to 11.56 mm/10 a for RCP4.5 and RCP8.5, respectively (Fig. 4). Seasonal precipitation presents a slightly increasing trend in spring and autumn, while a very slightly decreasing trend with the rate of 2.99 mm/10 a in summer for RCP4.5. On the other hand, the changing trends of winter precipitation for RCP8.5 is similar with that of RCP4.5 with a higher increasing rate. Notably, the summer precipitation of RCP8.5 shows an increase trend with the rate of 2.27 mm/10 a, which is opposite to that of RCP4.5.

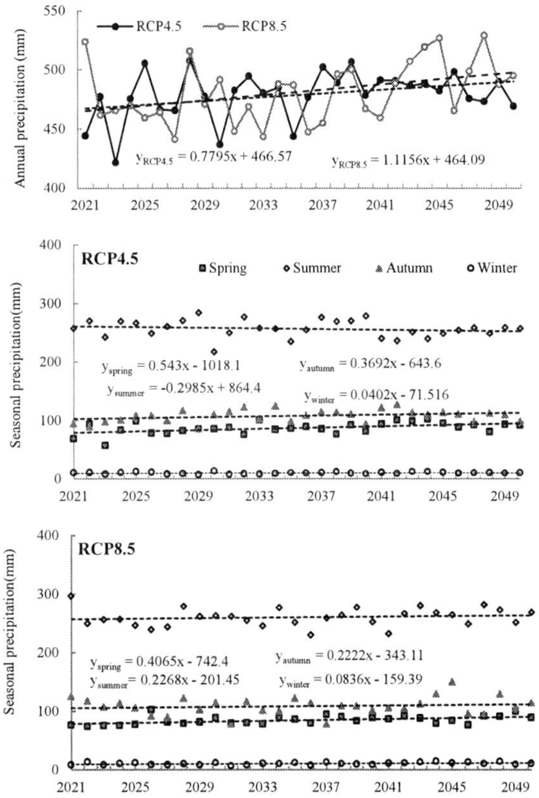

Figure 4. The time series of annual and seasonal precipitation for RCP4.5 and RCP8.5 from 2021 to 2050 in YRB.

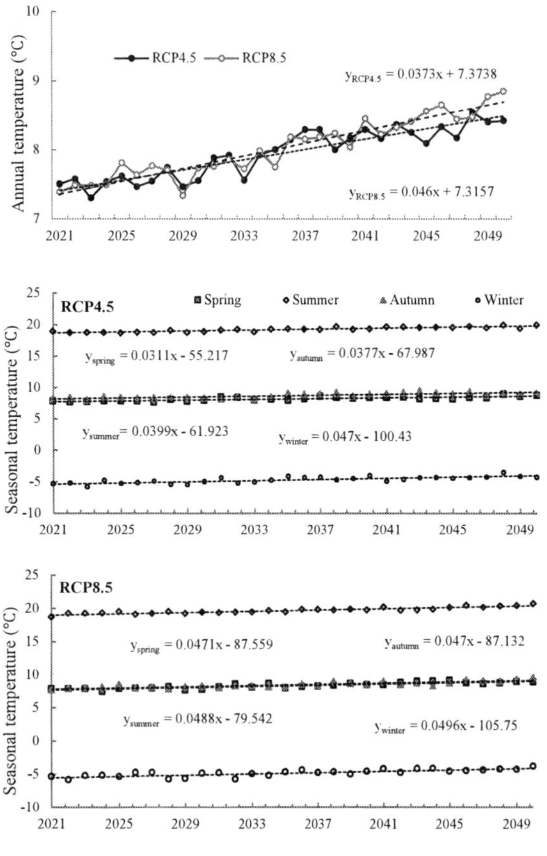

Figure 5. The time series of annual and seasonal temperature for RCP4.5 and RCP8.5 from 2021 to 2050 in YRB.

During the period of 2021–2050, the basin average temperature shows an obviously increasing for two RCPs (Fig. 5). Annual mean temperature projected an increasing rate of 0.373°/10 a and 0.46°/10 a for RCP4.5 and RCP8.5, respectively. Both RCP4.5 and RCP8.5 all project an increasing trend of seasonal temperature in YRB. The changing rate of seasonal temperature for RCP8.5 more than 0.47°/10 a, which is higher than that of RCP4. 5. Meanwhile, the fastest increasing of seasonal temperature occur in winter with the rate of 0.47°/10 a and 0.50°/10 a for RCP4.5 and RCP8.5, respectively. The changing rate of seasonal temperature for RCP4.5 is uneven for all seasons, in which autumn is increasing more quickly (about 0.07°/10 a) than that of spring. Whereas RCP8.5 projects a similar increasing rate in spring and autumn, while a little higher in summer and winter.

3.4 Ensemble streamflow prediction

During the period of 2021–2050, the spatial distribution of MK for annual streamflow for RCP4.5 scenario shows an obviously increasing trend in the middle of the basin, while decreasing in the west and north parts of basin. RCP8.5 projects a different trend of annual runoff with RCP4.5, which remarkable increasing in the east part of basin and decreasing in the west part of basin (Fig. 6). Seasonal streamflow of basin average for RCP4.5 and RCP8.5 show a slightly increasing trend except in summer of RCP4.5, which shows a slightly decreasing trend.

Notably, the spatial distribution of MK test for seasonal streamflow shows an uneven spatial changing trend (Fig. 7). For example, the streamflow shows a remarkable increasing trend in the sub-catchment of LZ in spring and autumn, while which shows a slightly decreasing trend in summer for RCP4.5. For the middle parts of YRB, the streamflow presents an unobvious decreasing trend in summer, while increasing in spring. For RCP8.5, streamflow of north parts of YRB shows a slightly increasing trend except the upstream region. Notably, the streamflow in upstream of YRB shows an obviously decreasing trend in summer and autumn for RCP8.5, while which shows an increasing trend in spring.

For further analysis of the changing trend of streamflow in YRB for the future period, we compared the changes of future streamflow with that of base period (1961–1990) in Fig. 8. In the future period, the VIC model simulated annual and seasonal streamflow reduce obviously compared with the base period. The annual streamflow will reduce about 10 % for two RCPs. The seasonal streamflow projected a reduce ratio of 3.44, 3.70, and 5.48 % in spring, summer, and autumn for RCP8.5, respectively, which is bigger than that of RCP4.5. Especially, the streamflow projected a remarkable reduce in winter about 41.11 and 35.11 % for RCP4.5 and RCP8.5, respectively.

4 Conclusions

The statistical downscaling method of EDCDFm is good at simulated historical period of precipitation and temperature in YRB. The eight models ensemble outputs show a higher correlation coefficient and lower RMSE than any single model. Projected precipitation and temperature all show an increasing trend during the period of 2021–2050. Compared to the base period (1961–2050), the annual streamflow of future period will reduced about 10 % and the streamflow in winter will reduce about 40 %. With the increasing of population and economic of YRB, there will be more require for the water resource in the future.

Data availability. The CMIP5 raw model outputs can be found in: https://esgf-node.llnl.gov/search/cmip5/. The observation data of weather stations in the YRB for the period of 1961–2005 can be found in China Meteorological Data Service Center (CMDC) (http://data.cma.cn/). Other data in this paper is not available.

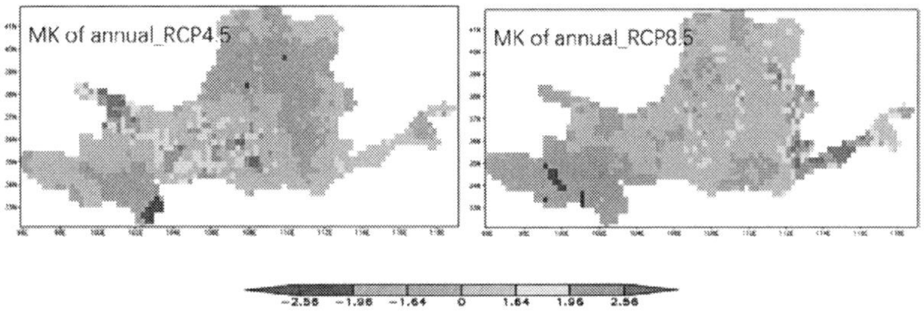

Figure 6. MK trend of annual streamflow for RCP4.5 and RCP8.5 from 2021 to 2050 in YRB.

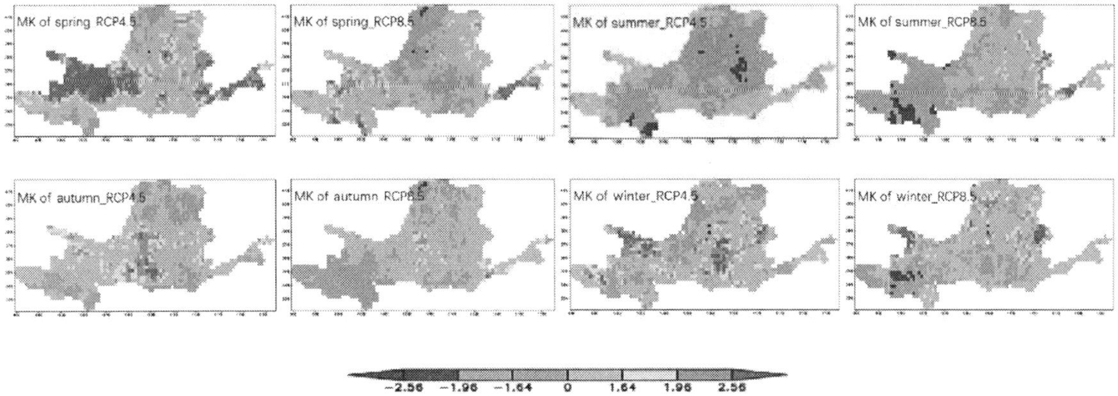

Figure 7. The spatial distribution of Mann-Kendall trend test of the streamflow during 2021 to 2050 in the YRB for two scenarios.

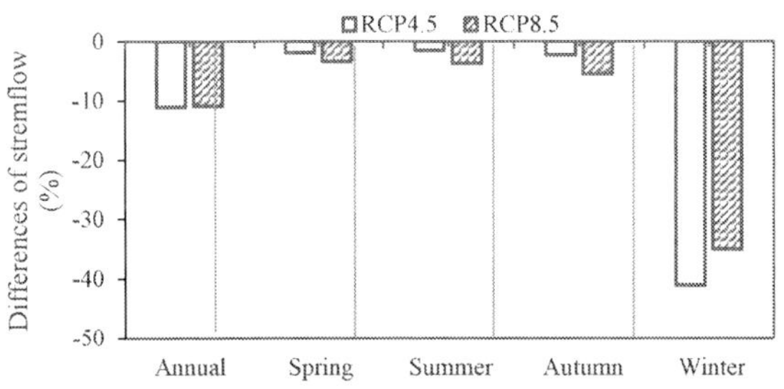

Figure 8. Projected changes of annual and seasonal streamflow (2021–2050) against to the base line period (1961–1990) for RCP4.5 and RCP8.5 in YRB.

Competing interests. The authors declare that they have no conflict of interest.

Special issue statement. This article is part of the special issue "Water security and the food–water–energy nexus: drivers, responses and feedbacks at local to global scales". It is a result of the IAHS Scientific Assembly 2017, Port Elizabeth, South Africa, 10–14 July 2017.

Acknowledgements. This research was financially supported by the National Key Research and Development Program under Grant No. 2016YFA0601504 approved by Ministry of Science and Technology, the People's Republic of China. This research has been also funded by The National Natural Science Foundation of China (51579066) and the Fundamental Research Funds for the Central Universities (2015B14514).

Edited by: Barry Croke
Reviewed by: three anonymous referees

References

Aloysius, N. R., Sheffield, J., Saiers, J. E., Li, H., and Wood, E. F.: Evaluation of historical and future simulations of precipitation and temperature in central Africa from CMIP5 climate models, J. Geophys. Res.-Atmos., 121, 130–152, 2016.

Duan, X. L., Hao, Z. C., Chen, Y., and Center, U.: Climate change forecasting in source region of yellow river based on bcsd downscaling method, Journal of Hohai University, 42, 195–199, 2014.

Hao, Z. C., Shi, F. X., and Wang, J. H.: Research and analysis of change of precipitation in headstream region of yellow river by using statistical downscaling model, Water Resources & Power, 29, 1–4, 2011.

Li, H., Sheffield, J., and Wood, E. F.: Bias correction of monthly precipitation and temperature fields from intergovernmental

panel on climate change ar4 models using equidistant quantile matching, J. Geophys. Res.-Atmos., 115, 985–993, 2010.
Murphy, J.: Predictions of climate change over Europe using statistical and dynamical downsaling techniques, Int. J. Climatol., 20, 489–501, 2000.
Nash, J. E. and Sutcliffe, J. V.: River flow forecasting through conceptual models part1 – a discussion of principles, J. Hydrol., 10, 282–290, 1970.
Stocker, T. F., Qin, D., Plattner, G.-K., Tignor, M., Allen, S. K., Boschung, J., Nauels, A., Xia, Y., Bex, V., and Midgley, P. M.: IPCC 2013: Summary for Policy Makers, Cambridge University Press, Cambridge, United Kingdom and New York, NY, USA, 2013.
Tareghian, R. and Rasmussen, P. F.: Statistical downscaling of precipitation using quantile regression, J. Hydrol., 487, 122–135, 2013.
Taylor, K. E.: Summarizing multiple aspects of model performance in a single diagram, J. Geophys. Res.-Atmos., 106, 7183–7192, 2001.
Wang, L. and Chen, W.: A CMIP5 multimodel projection of future temperature, precipitation, and climatological drought in china, Int. J. Climatol., 34, 2059–2078, 2014.
Wang, X. Y., Yang, T., and Hao, Z. C.: Climate extreme forecasting in headwater catchment of yellow river based on statistical downscaling method, Water Resources & Power, 29, 1–4, 2011.
Wilby, R. L., Dawson, C. W., and Barrow, E. M.: SDSM – a decision support tool for the assessment of regional climate change impacts, Environ. Modell. Softw., 17, 145–157, 2002.
Xu, Y., Gao, X., Giorgi, F.: Upgrades to the REA method for producing probabilistic climate change projections, Clim. Res., 41, 61–81, 2010.
Zhang, J. Y. and Wang, G. Q.: Study on the Impact of Climate Change on Hydrology and Water Resources, Science Press, Beijing, China, 2007.
Zhang, L. P., Chen, X. F., Zhang, X. L., and Song, X. Y.: A compare application research of VIC model and SWAT model in the mid-small valley flow simulation, Resources & Environment in the Yangtze Basin, 18, 745–752, 2009.

AUTHOR INDEX